THE GREAT MATHEMATICAL PROBLEMS

THE GREAT MATHEMATICAL PROBLEMS

Marvels and Mysteries of Mathematics

IAN STEWART

PROFILE BOOKS

First published in Great Britain in 2013 by
PROFILE BOOKS LTD
3A Exmouth House
Pine Street
London EC1R 0JH
www.profilebooks.com

1 3 5 7 9 10 8 6 4 2

Printed and bound in Great Britain by
Clays, Bungay, Suffolk

A CIP catalogue record for this book is available from the British Library.

ISBN 978 1 84668 1998
eISBN 978 184765 3512

FSC
Mixed Sources
Product group from well-managed
forests and other controlled sources
Cert no. SGS-COC-2061
www.fsc.org
© 1996 Forest Stewardship Council

The paper this book is printed on is certified by the © 1996 Forest
Stewardship Council A.C. (FSC). It is ancient-forest friendly. The printer
holds FSC chain of custody SGS-COC-2061

Contents

We must know. We shall know.
David Hilbert

Speech about mathematical problems in 1930, on the occasion of his honorary citizenship of Königsberg.[1]

Preface

athematics is a vast, ever-growing, ever-changing subject. Among the innumerable questions that mathematicians ask, and mostly answer, some stand out from the rest: prominent peaks that tower over the lowly foothills. These are the really big questions, the difficult and challenging problems that any mathematician would give his or her right arm to solve. Some remained unanswered for decades, some for centuries, a few for millennia. Some have yet to be conquered. Fermat's last theorem was an enigma for 350 years until Andrew Wiles dispatched it after seven years of toil. The Poincaré conjecture stayed open for over a century until it was solved by the eccentric genius Grigori Perelman, who declined all academic honours and a million-dollar prize for his work. The Riemann hypothesis continues to baffle the world's mathematicians, impenetrable as ever after 150 years.

The Great Mathematical Problems contains a selection of the really big questions that have driven the mathematical enterprise in radically new directions. It describes their origins, explains why they are important, and places them in the context of mathematics and science as a whole. It includes both solved and unsolved problems, which range over more than two thousand years of mathematical development, but its main focus is on questions that either remain open today, or have been solved within the past fifty years.

A basic aim of mathematics is to uncover the underlying simplicity of apparently complicated questions. This may not always be apparent, however, because the mathematician's conception of 'simple' relies on many technical and difficult concepts. An important feature of this book is to emphasise the deep simplicities, and avoid – or at the very least explain in straightforward terms – the complexities.

Mathematics is newer, and more diverse, than most of us imagine. At a rough estimate, the world's research mathematicians number about a hundred thousand, and they produce more than *two million* pages of new mathematics every year. Not 'new numbers', which are not what the enterprise is really about. Not 'new sums' like existing ones, but bigger – though we do work out some pretty big sums. One recent piece of algebra, carried out by a team of some 25 mathematicians, was described as 'a calculation the size of Manhattan'. That wasn't quite true, but it erred on the side of conservatism. The *answer* was the size of Manhattan; the calculation was a lot bigger. That's impressive, but what matters is quality, not quantity. The Manhattan-sized calculation qualifies on both counts, because it provides valuable basic information about a symmetry group that seems to be important in quantum physics, and is definitely important in mathematics. Brilliant mathematics can occupy one line, or an encyclopaedia – whatever the problem demands.

When we think of mathematics, what springs to mind is endless pages of dense symbols and formulas. However, those two million pages generally contain more words than symbols. The words are there to explain the background to the problem, the flow of the argument, the meaning of the calculations, and how it all fits into the ever-growing edifice of mathematics. As the great Carl Friedrich Gauss remarked around 1800, the essence of mathematics is 'notions, not notations'. Ideas, not symbols. Even so, the usual language for expressing mathematical ideas is symbolic. Many published research papers do contain more symbols than words. Formulas have a precision that words cannot always match.

However, it is often possible to explain the ideas while leaving out most of the symbols. *The Great Mathematical Problems* takes this as its guiding principle. It illuminates what mathematicians do, how they think, and why their subject is interesting and important. Significantly, it shows how today's mathematicians are rising to the challenges set by their predecessors, as one by one the great enigmas of the past surrender to the powerful techniques of the present, which changes the mathematics and science of the future. Mathematics ranks among humanity's greatest achievements, and its great problems – solved and unsolved – have guided and stimulated its astonishing power for millennia, both past and yet to come.

Coventry, June 2012

Figure Credits

1

Great problems

TELEVISION PROGRAMMES ABOUT MATHEMATICS are rare, good ones rarer. One of the best, in terms of audience involvement and interest as well as content, was Fermat's last theorem. The programme was produced by John Lynch for the British Broadcasting Corporation's flagship popular science series *Horizon* in 1996. Simon Singh, who was also involved in its making, turned the story into a spectacular bestselling book.[2] On a website, he pointed out that the programme's stunning success was a surprise:

> It was 50 minutes of mathematicians talking about mathematics, which is not the obvious recipe for a TV blockbuster, but the result was a programme that captured the public imagination and which received critical acclaim. The programme won the BAFTA for best documentary, a Priz Italia, other international prizes and an Emmy nomination – this proves that mathematics can be as emotional and as gripping as any other subject on the planet.

I think that there are several reasons for the success of both the television programme and the book and they have implications for the stories I want to tell here. To keep the discussion focused, I'll concentrate on the television documentary.

Fermat's last theorem is one of the truly great mathematical problems, arising from an apparently innocuous remark which one of the leading mathematicians of the seventeenth century wrote in the margin of a classic textbook. The problem became notorious because no one could prove what Pierre de Fermat's marginal note claimed, and this state of affairs continued for more than 300 years despite strenuous efforts by extraordinarily clever people. So when the British mathematician Andrew Wiles finally cracked the problem in 1995, the magnitude of his achievement was obvious to anyone. You didn't even need to know what the problem was, let alone how he had solved it. It was the mathematical equivalent of the first ascent of Mount Everest.

In addition to its significance for mathematics, Wiles's solution also involved a massive human-interest story. At the age of ten, he had become so intrigued by the problem that he decided to become a mathematician and solve it. He carried out the first part of the plan, and got as far as specialising in number theory, the general area to which Fermat's last theorem belongs. But the more he learned about real mathematics, the more impossible the whole enterprise seemed. Fermat's last theorem was a baffling curiosity, an isolated question of the kind that any number theorist could dream up without a shred of convincing evidence. It didn't fit into any powerful body of technique. In a letter to Heinrich Olbers, the great Gauss had dismissed it out of hand, saying that the problem had 'little interest for me, since a multitude of such propositions, which one can neither prove nor refute, can easily be formulated'.[3] Wiles decided that his childhood dream had been unrealistic and put Fermat on the back burner. But then, miraculously, other mathematicians suddenly made a breakthrough that linked the problem to a core topic in number theory, one on which Wiles was already an expert. Gauss, uncharacteristically, had underestimated the problem's significance, and was unaware that it could be linked to a deep, though apparently unrelated, area of mathematics.

With this link established, Wiles could now work on Fermat's enigma *and* do credible research in modern number theory at the same time. Better still, if Fermat didn't work out, anything significant that he discovered while trying to prove it would be publishable in its own right. So off the back burner it came, and Wiles began to think about Fermat's problem in earnest. After seven years of obsessive research, carried on in private and in secret – an unusual precaution in mathematics – he became convinced that he had found a solution. He delivered a series of lectures at a prestigious number theory conference, under an obscure title that fooled no one.[4] The exciting news broke, in the media as well as the halls of academe: Fermat's last theorem had been proved.

The proof was impressive and elegant, full of good ideas. Unfortunately, experts quickly discovered a serious gap in its logic. In attempts to demolish great unsolved problems of mathematics, this kind of development is depressingly common, and it almost always proves fatal. However, for once the Fates were kind. With assistance from his former student Richard Taylor, Wiles managed to bridge the gap, repair the proof, and complete his solution. The emotional burden involved became vividly clear in the television programme: it must have been the only occasion when a mathematician has burst into tears on screen, just recalling the traumatic events and the eventual triumph.

You may have noticed that I haven't told you what Fermat's last theorem *is*. That's deliberate; it will be dealt with in its proper place. As far as the success of the television programme goes, it doesn't actually matter. In fact, mathematicians have never greatly cared whether the theorem that Fermat scribbled in his margin is true or false, because nothing of great import hangs on the answer. So why all the fuss? Because a huge amount hangs on the inability of the mathematical community to *find* the answer. It's not just a blow to our self-esteem: it means that existing mathematical theories are missing something vital. In addition, the theorem is very easy to state; this adds to its air of mystery. How can something that seems so simple turn out to be so hard?

Although mathematicians didn't really care about the answer, they cared deeply that they didn't know what it was. And they cared even more about finding a method that could solve it, because that must surely shed light not just on Fermat's question, but on a host of others. This is often the case with great mathematical problems: it is the methods used to solve them, rather than the results themselves, that matter most. Of course, sometimes the actual result matters too: it depends on what its consequences are.

Wiles's solution is much too complicated and technical for television; in fact, the details are accessible only to specialists.[5] The proof does involve a nice mathematical story, as we'll see in due course, but any attempt to explain that on television would have lost most of the audience immediately. Instead, the programme sensibly concentrated on a more personal question: what is it like to tackle a notoriously difficult mathematical problem that carries a lot of historical baggage? Viewers were shown that there existed a small but dedicated band of mathematicians, scattered across the globe, who cared deeply about their research area, talked to each other, took note of each other's work, and devoted a large part of their lives to advancing mathematical knowledge. Their emotional investment and social interaction came over vividly. These were not clever automata, but real people, engaged with their subject. That was the message.

Those are three big reasons why the programme was such a success: a major problem, a hero with a wonderful human story, and a supporting cast of emotionally involved people. But I suspect there was a fourth, not quite so worthy. The majority of non-mathematicians seldom hear about new developments in the subject, for a variety of perfectly sensible reasons: they're not terribly interested anyway; newspapers hardly ever mention anything mathematical; when they do, it's often facetious or trivial; and nothing much in daily life seems to be affected by whatever it is that mathematicians are doing behind the scenes. All too often, school mathematics is presented as a closed book in which every question has an answer. Students can easily come to imagine that new mathematics is as rare as hen's teeth.

From this point of view, the big news was not that Fermat's last theorem had been proved. It was that at last *someone had done some new mathematics*. Since it had taken mathematicians more than 300 years to find a solution, many viewers subconsciously concluded that the breakthrough was the first important new mathematics discovered in the last 300 years. I'm not suggesting that they *explicitly* believed that. It ceases to be a sustainable position as soon as you ask some obvious questions, such as 'Why does the Government spend good money on university mathematics departments?' But subconsciously it was a common default assumption, unquestioned and unexamined. It made the magnitude of Wiles's achievement seem even greater.

One of the aims of this book is to show that mathematical research is thriving, with new discoveries being made all the time. You don't hear much about this activity because most of it is too technical for non-specialists, because most of the media are wary of anything intellectually more challenging than *The X Factor*, and because the applications of mathematics are deliberately hidden away to avoid causing alarm. 'What? My iPhone depends on advanced mathematics? How will I log in to Facebook when I failed my maths exams?'

Historically, new mathematics often arises from discoveries in other areas. When Isaac Newton worked out his laws of motion and his law of gravity, which together describe the motion of the planets, he did not polish off the problem of understanding the solar system. On the contrary, mathematicians had to grapple with a whole new range of questions: yes, we know the laws, but what do they imply? Newton invented calculus to answer that question, but his new method also has limitations. Often it rephrases the question instead of providing the answer. It turns the problem into a special kind of formula, called a differential equation, whose *solution* is the answer. But you still have to solve the equation. Nevertheless, calculus was a brilliant start. It showed us that answers were possible, and it provided one effective way to seek them, which continues to provide major insights more than 300 years later.

As humanity's collective mathematical knowledge grew, a second source of inspiration started to play an increasing role in the creation of even more: the internal demands of mathematics itself. If, for example, you know how to solve algebraic equations of the first, second, third, and fourth degree, then you don't need much imagination to ask about the fifth degree. (The degree is basically a measure of complexity, but you don't even need to know what it is to ask the obvious question.) If a solution proves elusive, as it did, that fact *alone* makes mathematicians even more determined to find an answer, whether or not the result has useful applications.

I'm not suggesting applications don't matter. But if a particular piece of mathematics keeps appearing in questions about the physics of waves – ocean waves, vibrations, sound, light – then it surely makes sense to investigate the gadget concerned in its own right. You don't need to know ahead of time exactly how any new idea will be used: the topic of waves is common to so many important areas that significant new insights are bound to be useful for something. In this case, those somethings included radio, television, and radar.[6] If somebody thinks up a new way to understand heat flow, and comes up with a brilliant new technique that unfortunately lacks proper mathematical support, then it makes sense to sort the whole thing out *as a piece of mathematics*. Even if you don't give a fig about how heat flows, the results might well be applicable elsewhere. Fourier analysis, which emerged from this particular line of investigation, is arguably the most useful single mathematical idea ever found. It underpins modern telecommunications, makes digital cameras possible, helps to clean up old movies and recordings, and a modern extension is used by the FBI to store fingerprint records.[7]

After a few thousand years of this kind of interchange between the external uses of mathematics and its internal structure, these two aspects of the subject have become so densely interwoven that picking them apart is almost impossible. The mental attitudes involved are more readily distinguishable, though, leading to a broad classification of mathematics into two kinds: pure and applied. This is defensible as a rough-and-ready way to locate mathematical ideas in the intellectual landscape, but it's not a terribly accurate description of the subject itself. At best it distinguishes two ends of a continuous spectrum of mathematical styles. At worst, it misrepresents which parts of the subject are useful and where the ideas come from. As with all branches of science, what gives mathematics its power is the *combination* of abstract reasoning and inspiration from the outside world, each feeding off the other. Not only is it impossible to pick the two strands apart: it's pointless.

Most of the really important mathematical problems, the great problems that this book is about, have arisen within the subject through a kind of intellectual navel-gazing. The reason is simple: they are *mathematical* problems. Mathematics often looks like a collection of isolated areas, each with its own special techniques: algebra, geometry, trigonometry, analysis, combinatorics, probability. It tends to be taught that way, with good reason: locating each separate topic in a single well-defined area helps students to organise the material in their minds. It's a reasonable first approximation to the structure of mathematics, especially long-established mathematics. At the research frontiers, however, this tidy delineation often breaks down. It's not just that the boundaries between the major areas of mathematics are blurred. It's that they don't really exist.

Every research mathematician is aware that, at any moment, suddenly and unpredictably, the problem they are working on may turn out to require ideas from some apparently unrelated area. Indeed, new research often combines areas. For instance, my own research mostly centres on pattern formation in dynamical systems, systems that change over time according to specific rules. A typical example is the way animals move. A trotting horse repeats the same sequence of leg movements over and over again, and there is a clear pattern: the legs hit the ground together in diagonally related pairs. That is, first the front left and back right legs hit, then the other two. Is this a problem about patterns, in which case the appropriate methods come from group theory, the algebra of symmetry? Or is it a problem about dynamics, in which case the appropriate area is Newtonian-style differential equations?

The answer is that, by definition, it has to be both. It is not their intersection, which would be the material they have in common – basically, nothing. Instead, it is a new 'area', which straddles two of the traditional divisions of mathematics. It is like a bridge across a river that separates two countries; it links the two, but belongs to neither. But this bridge is not a thin strip of roadway; it is comparable in size to each of the countries. Even more vitally, the methods involved are not limited to those two areas. In fact, virtually every course in mathematics that I have ever studied has played a role somewhere in my research. My Galois theory course as an undergraduate at Cambridge was about how to solve (more precisely, why we can't solve) an algebraic equation of the fifth degree. My graph theory course was about networks, dots joined by lines. I never took a course in dynamical systems, because my PhD was in algebra, but over the years I picked up the basics, from steady states to chaos. Galois theory, graph theory, dynamical systems: three separate areas. Or so I assumed until 2011, when I wanted to understand how to detect chaotic dynamics in a network of dynamical systems, and a crucial step depended on things I'd learned 45 years earlier in my Galois theory course.

Mathematics, then, is not like a political map of the world, with each speciality neatly surrounded by a clear boundary, each country tidily distinguished from its neighbours by being coloured pink, green, or pale blue. It is more like a natural landscape, where you can never really say where the valley ends and the foothills begin, where the forest merges into woodland, scrub, and grassy plains, where lakes insert regions of water into every other kind of terrain, where rivers link the snow-clad slopes of the mountains to the distant, low-lying oceans. But this ever-changing mathematical landscape consists not of rocks, water, and plants, but of ideas; it is tied together not by geography, but by logic. And it is a dynamic landscape, which changes as new ideas and methods are discovered or invented. Important concepts with extensive implications are like mountain peaks, techniques with lots of uses are like broad rivers that carry travellers across the fertile plains. The more clearly defined the landscape becomes, the easier it is to spot unscaled peaks, or unexplored terrain that creates unwanted obstacles. Over time, some of the peaks and obstacles acquire iconic status. These are the great problems.

What makes a great mathematical problem great? Intellectual depth, combined with simplicity and elegance. Plus: it has to be *hard*. Anyone can climb a hillock; Everest is another matter entirely. A great problem is usually simple to state, although the terms required may be elementary or highly technical. The statements of Fermat's last theorem and the four colour problem make immediate sense to anyone familiar with school mathematics. In contrast, it is impossible even to state the Hodge conjecture or the mass gap hypothesis without invoking deep concepts at the research frontiers – the latter, after all, comes from quantum field theory. However, to those versed in such areas, the statement of the question concerned is simple and natural. It does not involve pages and pages of dense, impenetrable text. In between are problems that require something at the level of undergraduate mathematics, if you want to understand them in complete detail. A more general feeling for the essentials of the problem – where it came from, why it's important, what you could do if you possessed a solution – is usually accessible to any interested person, and that's what I will be attempting to provide. I admit that the Hodge conjecture is a hard nut to crack in that respect, because it is very technical and very abstract. However, it is one of the seven Clay Institute millennium mathematics problems, with a million-dollar prize attached, and it absolutely must be included.

Great problems are creative: they help to bring new mathematics into being. In 1900 David Hilbert delivered a lecture at the

International Congress of Mathematicians in Paris, in which he listed 23 of the most important problems in mathematics. He didn't include Fermat's last theorem, but he mentioned it in his introduction. When a distinguished mathematician lists what he thinks are some of the great problems, other mathematicians pay attention. The problems wouldn't be on the list unless they were important, and hard. It is natural to rise to the challenge, and try to answer them. Ever since, solving one of Hilbert's problems has been a good way to win your mathematical spurs. Many of these problems are too technical to include here, many are open-ended programmes rather than specific problems, and several appear later in their own right. But they deserve to be mentioned, so I've put a brief summary in the notes.[8]

That's what makes a great mathematical problem great. What makes it problematic is seldom deciding what the answer should be. For virtually all great problems, mathematicians have a very clear idea of what the answer ought to be – or had one, if a solution is now known. Indeed, the statement of the problem often includes the expected answer. Anything described as a conjecture is like that: a plausible guess, based on a variety of evidence. Most well-studied conjectures eventually turn out to be correct, though not all. Older terms like hypothesis carry the same meaning, and in the Fermat case the word 'theorem' is (more precisely, was) abused – a theorem requires a proof, but that was precisely what was missing until Wiles came along.

Proof, in fact, is the requirement that makes great problems problematic. Anyone moderately competent can carry out a few calculations, spot an apparent pattern, and distil its essence into a pithy statement. Mathematicians demand more evidence than that: they insist on a complete, logically impeccable proof. Or, if the answer turns out to be negative, a disproof. It isn't really possible to appreciate the seductive allure of a great problem without appreciating the vital role of proof in the mathematical enterprise. Anyone can make an educated guess. What's hard is to prove it's right. Or wrong.

The concept of mathematical proof has changed over the course of history, with the logical requirements generally becoming more stringent. There have been many highbrow philosophical discussions of the nature of proof, and these have raised some important issues. Precise logical definitions of 'proof' have been proposed and implemented. The one we teach to undergraduates is that a proof begins with a collection of explicit assumptions called axioms. The axioms are, so to speak, the rules of the game. Other axioms are possible, but they lead to different games. It was Euclid, the ancient Greek geometer, who introduced this approach to mathematics, and it is still valid today. Having agreed on the axioms, a proof of some statement is a series of steps, each of which is a logical consequence of either the axioms, or previously proved statements, or both. In effect, the mathematician is exploring a logical maze, whose junctions are statements and whose passages are valid deductions. A proof is a path through the maze, starting from the axioms. What it proves is the statement at which it terminates.

However, this tidy concept of proof is not the whole story. It's not even the most important part of the story. It's like saying that a symphony is a sequence of musical notes, subject to the rules of harmony. It misses out all of the creativity. It doesn't tell us how to find proofs, or even how to validate other people's proofs. It doesn't tell us which locations in the maze are significant. It doesn't tell us which paths are elegant and which are ugly, which are important and which are irrelevant. It is a formal, mechanical description of a process that has many other aspects, notably a human dimension. Proofs are discovered by people, and research in mathematics is not just a matter of step-by-step logic.

Taking the formal definition of proof literally can lead to proofs that are virtually unreadable, because most of the time is spent dotting logical i's and crossing logical t's in circumstances where the outcome already stares you in the face. So practising mathematicians cut to the chase, and leave out anything that is routine or obvious. They make it clear that there's a gap by using stock phrases like 'it is easy to verify that' or 'routine calculations imply'. What they don't do, at least not consciously, is to slither past a logical difficulty and to try to pretend it's not there. In fact, a competent mathematician will go out of his or her way to point out exactly those parts of the argument that are logically fragile, and they will devote most of their time to explaining how to make them sufficiently robust. The upshot is that a proof, in practice, is a mathematical story with its own narrative flow. It has a beginning, a middle, and an end. It often has subplots, growing out of the main plot, each with its own resolution. The British mathematician

Christopher Zeeman once remarked that a theorem is an intellectual resting point. You can stop, get your breath back, and feel you've got somewhere definite. The subplot ties off a loose end in the main story. Proofs resemble narratives in other ways: they often have one or more central characters – ideas rather than people, of course – whose complex interactions lead to the final revelation.

As the undergraduate definition indicates, a proof starts with some clearly stated assumptions, derives logical consequences in a coherent and structured way, and ends with whatever it is you want to prove. But a proof is not just a list of deductions, and logic is not the sole criterion. A proof is a story told to and dissected by people who have spent much of their life learning how to read such stories and find mistakes or inconsistencies: people whose main aim is to prove the storyteller *wrong*, and who possess the uncanny knack of spotting weaknesses and hammering away at them until they collapse in a cloud of dust. If any mathematician claims to have solved a significant problem, be it a great one or something worthy but less exalted, the professional reflex is not to shout 'hurray!' and sink a bottle of champagne, but to try to shoot it down.

That may sound negative, but proof is the only reliable tool that mathematicians have for making sure that what they say is correct. Anticipating this kind of response, researchers spend a lot of their effort trying to shoot their own ideas and proofs down. It's less embarrassing that way. When the story has survived this kind of critical appraisal, the consensus soon switches to agreement that it is correct, and at that point the inventor of the proof receives appropriate praise, credit, and reward. At any rate, that's how it usually works out, though it may not always seem that way to the people involved. If you're close to the action, your picture of what's going on may be different from that of a more detached observer.

How do mathematicians solve problems? There have been few rigorous scientific studies of this question. Modern educational research, based on cognitive science, largely focuses on education up to high school level. Some studies address the teaching of undergraduate mathematics, but those are relatively few. There are significant differences between learning and teaching existing mathematics and creating new mathematics. Many of us can play a musical instrument, but far fewer can compose a concerto or even write a pop song.

When it comes to creativity at the highest levels, much of what we know – or think we know – comes from introspection. We ask mathematicians to explain their thought processes, and seek general principles. One of the first serious attempts to find out how mathematicians think was Jacques Hadamard's *The Psychology of Invention in the Mathematical Field*, first published in 1945.[9] Hadamard interviewed leading mathematicians and scientists of his day and asked them to describe how they thought when working on difficult problems. What emerged, very strongly, was the vital role of what for lack of a better term must be described as intuition. Some feature of the subconscious mind guided their thoughts. Their most creative insights did not arise through step by step logic, but by sudden, wild leaps.

One of the most detailed descriptions of this apparently illogical approach to logical questions was provided by the French mathematician Henri Poincaré, one of the leading figures of the late nineteenth and early twentieth centuries. Poincaré ranged across most of mathematics, founding several new areas and radically changing many others. He plays a prominent role in several later chapters. He also wrote popular science books, and this breadth of experience may have helped him to gain a deeper understanding of his own thought processes. At any rate, Poincaré was adamant that conscious logic was only part of the creative process. Yes, there were times when it was indispensable: deciding what the problem really was, systematically verifying the answer. But in between, Poincaré felt that his brain was often working on the problem without telling him, in ways that he simply could not fathom.

His outline of the creative process distinguished three key stages: preparation, incubation, and illumination. Preparation consists of conscious logical efforts to pin the problem down, make it precise, and attack it by conventional methods. This stage Poincaré considered essential: it gets the subconscious going and provides raw materials for it to work with. Incubation takes place when you stop thinking about the problem and go off and do something else. The subconscious now starts combining ideas with each other, often quite wild ideas, until light starts to dawn. With luck, this leads to illumination: your subconscious taps you on the shoulder and the proverbial light bulb goes off in your mind.

This kind of creativity is like walking a tightrope. On the one hand, you won't solve a difficult problem unless you make yourself familiar with the area to which it seems to belong – along with many other areas, which may or may not be related, just in case they are. On the other hand, if all you do is get trapped into standard ways of thinking, which others have already tried, fruitlessly, then you will be stuck in a mental rut and discover nothing new. So the trick is to know a lot, integrate it consciously, put your brain in gear for weeks ... and then set the question aside. The intuitive part of your mind then goes to work, rubs ideas against each other to see whether the sparks fly, and notifies you when it has found something. This can happen at any moment: Poincaré suddenly saw how to solve a problem that had been bugging him for months when he was stepping off a bus. Srinivasa Ramanujan, a self-taught Indian mathematician with a talent for remarkable formulas, often got his ideas in dreams. Archimedes famously worked out how to test metal to see if it were gold when he was having a bath.

Poincaré took pains to point out that without the initial period of preparation, progress is unlikely. The subconscious, he insisted, needs to be given plenty to think about, otherwise the fortuitous combinations of ideas that will eventually lead to a solution cannot form. Perspiration begets inspiration. He must also have known – because any creative mathematician does – that this simple three-stage process seldom occurs just once. Solving a problem often requires more than one breakthrough. The incubation stage for one idea may be interrupted by a subsidiary process of preparation, incubation, and illumination for something that is needed to make the first idea work. The solution to any problem worth its salt, be it great or not, typically involves many such sequences, nested inside each other like one of Benoît Mandelbrot's intricate fractals. You solve a problem by breaking it down into subproblems. You convince yourself that if you can solve these subproblems, then you can assemble the results to solve the whole thing. Then you work on the subproblems. Sometimes you solve one; sometimes you fail, and a rethink is in order. Sometimes a subproblem itself breaks up into more pieces. It can be quite a task just to keep track of the plan.

I described the workings of the subconscious as 'intuition'. This is one of those seductive words like 'instinct', which is widely used even though it is devoid of any real meaning. It's a name for something whose presence we recognise, but which we do not understand. Mathematical intuition is the mind's ability to sense form and structure, to detect patterns that we cannot consciously perceive. Intuition lacks the crystal clarity of conscious logic, but it makes up for that by drawing attention to things we would never have consciously considered. Neuroscientists are barely starting to understand how the brain carries out much simpler tasks. But however intuition works, it must be a consequence of the structure of the brain and how it interacts with the external world.

Often the key contribution of intuition is to make us aware of weak points in a problem, places where it may be vulnerable to attack. A mathematical proof is like a battle, or if you prefer a less warlike metaphor, a game of chess. Once a potential weak point has been identified, the mathematician's technical grasp of the machinery of mathematics can be brought to bear to exploit it. Like Archimedes, who wanted a firm place to stand so that he could move the Earth, the research mathematician needs some way to exert leverage on the problem. One key idea can open it up, making it vulnerable to standard methods. After that, it's just a matter of technique.

My favourite example of this kind of leverage is a puzzle that has no intrinsic mathematical significance, but drives home an important message. Suppose you have a chessboard, with 64 squares, and a supply of dominoes just the right size to cover two adjacent squares of the board. Then it's easy to cover the entire board with 32 dominoes. But now suppose that two diagonally opposite corners of the board have been removed, as in Figure 1. Can the remaining 62 squares be covered using 31 dominoes? If you experiment, nothing seems to work. On the other hand, it's hard to see any obvious reason for the task to be impossible. Until you realise that however the dominoes are arranged, each of them must cover one black square and one white square. This is your lever; all you have to do now is to wield it. It implies that any region covered by dominoes contains the same number of black squares as it does white squares. But diagonally opposite

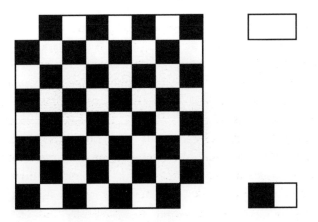

Fig 1 Can you cover the hacked chessboard with dominoes, each covering two squares (top right)? If you colour the domino (bottom right) and count how many black and white squares there are, the answer is clear.

squares have the same colour, so removing two of them (here white ones) leads to a shape with two more black squares than white. So no such shape can be covered. The observation about the combination of colours that *any* domino covers is the weak point in the puzzle. It gives you a place to plant your logical lever, and push. If you were a medieval baron assaulting a castle, this would be the weak point in the wall – the place where you should concentrate the firepower of your trebuchets, or dig a tunnel to undermine it.

Mathematical research differs from a battle in one important way. Any territory you once occupy remains yours for ever. You may decide to concentrate your efforts somewhere else, but once a theorem is proved, it doesn't disappear again. This is how mathematicians make progress on a problem, even when they fail to solve it. They establish a new fact, which is then available for anyone else to use, in any context whatsoever. Often the launchpad for a fresh assault on an age-old problem emerges from a previously unnoticed jewel half-buried in a shapeless heap of assorted facts. And that's one reason why new mathematics can be important for its own sake, even if its uses are not immediately apparent. It is one more piece of territory occupied, one more weapon in the armoury. Its time may yet come – but it certainly won't if it is deemed 'useless' and forgotten, or never allowed to come into existence because no one can see what it is *for*.

2

Prime territory
Goldbach Conjecture

S OME OF THE GREAT PROBLEMS show up very early in our mathematical education, although we may not notice. Soon after we are taught multiplication, we run into the concept of a prime number. Some numbers can be obtained by multiplying two smaller numbers together; for example, $6 = 2 \times 3$. Others, such as 5, cannot be broken up in this manner; the best we can do is $5 = 1 \times 5$, which doesn't involve two *smaller* numbers. Numbers that can be broken up are said to be composite; those that can't are prime. Prime numbers seem such simple things. As soon as you can multiply whole numbers together you can understand what a prime number is. Primes are the basic building blocks for whole numbers, and they turn up all over mathematics. They are also deeply mysterious, and they seem to be scattered almost at random. There's no doubting it: primes are an enigma. Perhaps this is a consequence of their definition – not so much what they are as what they are not. On the other hand, they are fundamental to mathematics, so we can't just throw up our hands in horror and give up. We need to come to terms with primes, and ferret out their innermost secrets.

A few features are obvious. With the exception of the smallest prime, 2, all primes are odd. With the exception of 3, the sum of their digits can't be a multiple of 3. With the exception of 5, they can't end in the digit 5. Aside from these rules, and a few subtler ones, you can't look at a number and immediately spot whether it is prime. There do exist formulas for primes, but to a great extent they are cheats: they don't provide useful new information about primes; they are just clever ways to encode the definition of 'prime' in a formula. Primes are like people: they are individuals, and they don't conform to standard rules.

Over the millennia, mathematicians have gradually increased their understanding of prime numbers, and every so often another big problem about them is solved. However, many questions still remain unanswered. Some are basic and easy to state; others are more esoteric. This chapter discusses what we do and don't know about these infuriating, yet fundamental, numbers. It begins by setting up some of the basic concepts, in particular, prime factorisation – how to express a given number by multiplying primes together. Even this familiar process leads into deep waters as soon as we start asking for genuinely effective methods for finding a number's prime factors. One surprise is that it seems to be relatively easy to test a number to determine whether it is prime, but if it's composite, finding its prime factors is often much harder.

Having sorted out the basics, we move on to the most famous unsolved problem about primes, the 250-year-old Goldbach conjecture. Recent progress on this question has been dramatic, but not yet decisive. A few other problems provide a brief sample of what is still to be discovered about this rich but unruly area of mathematics.

Prime numbers and factorisation are familiar from school arithmetic, but most of the interesting features of primes are seldom taught at that level, and virtually nothing is proved. There are sound reasons for that: the proofs, even of apparently obvious properties, are surprisingly hard. Instead, pupils are taught some simple methods for working with primes, and the emphasis is on calculations with relatively small numbers. As a result, our early experience of primes is a bit misleading.

The ancient Greeks knew some of the basic properties of primes, and they knew how to prove them. Primes and factors are the main topic of Book VII of Euclid's *Elements*, the great geometry classic. This particular book contains a geometric presentation of division and multiplication in arithmetic. The Greeks preferred to work with lengths of lines, rather than numbers as such, but it is easy to reformulate their results in the language of numbers. Euclid takes care to prove statements that may seem obvious: for example, Proposition

16 of Book VII proves that when two numbers are multiplied together, the result is independent of the order in which they are taken. That is, $ab = ba$, a basic law of algebra.

In school arithmetic, prime factors are used to find the greatest common divisor (or highest common factor) of two numbers. For instance, to find the greatest common divisor of 135 and 630, we factorise them into primes:

$$135 = 3^3 \times 5 \qquad 630 = 2 \times 3^2 \times 5 \times 7$$

Then, for each prime, we take the largest power that occurs in both factorisations, obtaining $3^2 \times 5$. Multiply out to get 45: this is the greatest common divisor. This procedure gives the impression that prime factorisation is needed to find greatest common divisors. Actually, the logical relationship goes the other way. Book VII Proposition 2 of the *Elements* presents a method for finding the greatest common divisor of two whole numbers without factorising them. It works by repeatedly subtracting the smaller number from the larger one, then applying a similar process to the resulting remainder and the smaller number, and continuing until there is no remainder. For 135 and 630, a typical example using smallish numbers, the process goes like this. Subtract 135 repeatedly from 630:

$$630 - 135 = 495$$
$$495 - 135 = 360$$
$$360 - 135 = 225$$
$$225 - 135 = 90$$

Since 90 is smaller than 135, switch to the two numbers 90 and 135:

$$135 - 90 = 45$$

Since 45 is smaller than 90, switch to 45 and 90:

$$90 - 45 = 45$$
$$45 - 45 = 0$$

Therefore the greatest common divisor of 135 and 630 is 45.

This procedure works because at each stage it replaces the original pair of numbers by a simpler pair (one of the numbers is smaller) that has the same greatest common divisor. Eventually one of the numbers divides the other exactly, and at that stage we stop. Today's term for an explicit computational method that is guaranteed to find an answer to a given problem is 'algorithm'. So Euclid's procedure is now called the Euclidean algorithm. It is logically prior to prime factorisation. Indeed, Euclid uses his algorithm to prove basic properties about prime factors, and so do university courses in mathematics today.

Euclid's Proposition 30 is vital to the whole enterprise. In modern terms, it states that if a prime divides the product of two numbers – what you get by multiplying them together – then it must divide one of them. Proposition 32 states that either a number is prime or it has a prime factor. Putting the two together, it is easy to deduce that every number is a product of prime factors, and that this expression is unique apart from the order in which the factors are written. For example,

$$60 = 2 \times 2 \times 3 \times 5 = 2 \times 3 \times 2 \times 5 = 5 \times 3 \times 2 \times 2$$

and so on, but the only way to get 60 is to rearrange the first factorisation. There is no factorisation, for example, looking like $60 = 7 \times$ *something*. The existence of the factorisation comes from Proposition 32. If the number is prime, stop. If not, find a prime factor, divide to get a smaller number, and repeat. Uniqueness comes from Proposition 30. For example, if there were a factorisation $60 = 7 \times$ *something*, then 7 must divide one of the numbers 2, 3, or 5, but it doesn't.

At this point I need to clear up a small but important point: the exceptional status of the number 1. According to the definition as stated so far, 1 is clearly prime: if we try to break it up, the best we can do is $1 = 1 \times 1$, which does not involve smaller numbers. However, this interpretation causes problems later in the theory, so for the last century or two, mathematicians have added an extra restriction. The number 1 is so special that it should be considered as neither prime nor composite. Instead, it is a third manner of beast, a unit. One reason for treating 1 as a special case, rather than a genuine prime, is that if we call 1 a prime then uniqueness fails. In fact, $1 \times 1 = 1$ already exhibits the failure, and $1 \times 1 \times 1 \times 1 \times 1 \times 1 \times 1 \times 1 = 1$ rubs our noses in it. We could modify uniqueness to say 'unique except for extra 1s', but that's just another way to admit that 1 is special.

Much later, in Proposition 20 of Book IX, Euclid proves another key fact: 'Prime numbers are more than any assigned multitude of prime numbers.' That is, the number of primes is infinite. It's a wonderful theorem with a clever proof, but it opened up a huge can of worms. If the primes go on for ever, yet seem to have no pattern, how can we describe what they look like?

We have to face up to that question because we can't ignore the primes. They are essential features of the mathematical landscape. They are especially common, and useful, in number theory. This area of mathematics studies properties of whole numbers. That may sound a bit elementary, but actually number theory is one of the deepest and most difficult areas of mathematics. We will see plenty of evidence for that statement later. In 1801 Gauss, the leading number theorist of his age – arguably one of the leading mathematicians of all time, perhaps even the greatest of them all – wrote an advanced textbook of number theory, the *Disquisitiones Arithmeticae* ('Investigations in arithmetic'). In among the high-level topics, he pointed out that we should not lose sight of two very basic issues: 'The problem of distinguishing prime numbers from composite numbers and of resolving the latter into their prime factors is known to be one of the most important and useful in arithmetic.'

At school, we are usually taught exactly one way to find the prime factors of a number: try all possible factors in turn until you find something that goes exactly. If you haven't found a factor by the time you reach the square root of the original number – more precisely, the largest whole number that is less than or equal to that square root – then the number is prime. Otherwise you find a factor, divide out by that, and repeat. It's more efficient to try just prime factors, which requires having a list of primes. You stop at the square root because the smallest factor of any composite number is no greater than its square root. However, this procedure is hopelessly inefficient when the numbers become large. For example, if the number is

$$1,080,813,321,843,836,712,253$$

then its prime factorisation is

$$13,929,010,429 \times 77,594,408,257$$

and you would have to try the first 624,401,249 primes in turn to find the smaller of the two factors. Of course, with a computer this is fairly easy, but if we start with a 100-digit number that happens to be the product of two 50-digit numbers, and employ a systematic search through successive primes, the universe will end before the computer finds the answer.

In fact, today's computers can generally factorise 100-digit numbers. My computer takes less than a second to find the prime factors of $10^{99} + 1$, which looks like 1000 ... 001 with 98 zeros. It is a product of 13 primes (one of them occurs twice), of which the smallest is 7 and the largest is

$$141,122,524,877,886,182,282,233,539,317,796,144,938,305,$$
$$111,168,717$$

But if I tell the computer to factorise $10^{199} + 1$, with 200 digits, it churns away for ages and gets nowhere. Even so, the 100-digit calculation is impressive. What's the secret? Find more efficient methods than trying all potential prime factors in turn.

We now know a lot more than Gauss did about the first of his problems (testing for primes) and a lot less than we'd like to about the second (factorisation). The conventional wisdom is that primality testing is far simpler than factorisation. This generally comes as a surprise to non-mathematicians, who were taught at school to test whether a number is prime by the same method used for factorisation: try all possible divisors. It turns out that there are slick ways to prove that a number is prime without doing that. They also allow us to prove that a number is composite, without finding any of its factors. Just show that it fails a primality test.

The great grand-daddy of all modern primality tests is Fermat's theorem, not to be confused with the celebrated Fermat's last theorem, chapter 7. This theorem is based on modular arithmetic, sometimes known as 'clock arithmetic' because the numbers wrap round like those on a clock face. Pick a number – for a 12-hour analogue clock it is 12 – and call it the modulus. In any arithmetical calculation with whole numbers, you now allow yourself to replace any multiple of 12 by zero. For example, $5 \times 5 = 25$, but 24 is twice 12, so subtracting 24 we obtain $5 \times 5 = 1$ to the modulus 12. Modular arithmetic is very pretty, because nearly all of the usual rules of arithmetic still work. The main difference is that you can't always divide one number by another, even when it's not zero. Modular arithmetic is also useful, because it provides a tidy way to deal with questions about divisibility: which numbers are divisible by the chosen modulus, and what is the remainder when they're not? Gauss introduced modular arithmetic in the *Disquisitiones Arithmeticae*, and today it is widely used in computer science, physics, and engineering, as well as mathematics.

Fermat's theorem states that if we choose a prime modulus p, and take any number a that is not a multiple of p, then the $(p-1)$ th power of a is equal to 1 in arithmetic to the modulus p. Suppose, for example, that $p = 17$ and $a = 3$. Then the theorem predicts that when we divide 3^{16} by 17, the remainder is 1. As a check,

$$3^{16} = 43,046,721 = 2,532,160 \times 17 + 1$$

No one in their right mind would want to do the sums that way for, say, 100-digit primes. Fortunately, there is a clever, quick way to carry out this kind of calculation. The point is that if the answer is not equal to 1 then the modulus we started with is composite. So Fermat's theorem forms the basis of an efficient test that provides a necessary condition for a number to be prime.

Unfortunately, the test is not sufficient. Many composite numbers, known as Carmichael numbers, pass the test. The smallest is 561, and in 2003 Red Alford, Andrew Granville, and Carl Pomerance proved, to general amazement, that there are infinitely many. The amazement was because they found a proof; the actual result was less of a surprise. In fact, they showed that there are at least $x^{2/7}$ Carmichael numbers less than or equal to x if x is large enough.

However, more sophisticated variants of Fermat's theorem can be turned into genuine tests for primality, such as one published in 1976 by Gary Miller. Unfortunately, the proof of the validity of Miller's test depends on an unsolved great problem, the generalised Riemann hypothesis, chapter 9. In 1980 Michael Rabin turned Miller's test into a probabilistic one, a test that might occasionally give the wrong answer. The exceptions, if they exist, are very rare, but they can't be ruled out altogether. The most efficient deterministic (that is, guaranteed correct) test to date is the Adleman-Pomerance-Rumely test, named for Leonard Adleman, Pomerance, and Robert Rumely. It uses ideas from number theory that are more sophisticated than Fermat's theorem, but in a similar spirit.

I still vividly recall a letter from one hopeful amateur, who proposed a variant of trial division. Try all possible divisors, but start at the square root and work *downwards*. This method sometimes gets the answer more quickly than doing things in the usual order, but as the numbers get bigger it runs into the same kind of trouble as the usual method. If you try it on my example above, the 22-digit number 1,080,813,321,843,836,712,253, then the square root is about 32,875,725,419. You have to try 794,582,971 prime divisors before you find one that works. This is *worse* than searching in the usual direction.

In 1956 The famous logician Kurt Gödel, writing to John von Neumann, echoed Gauss's plea. He asked whether trial division could be improved, and if so, by how much. Von Neumann didn't pursue the question, but over the years others answered Gödel by discovering practical methods for finding primes with up to 100 digits, sometimes more. These methods, of which the best known is called the quadratic sieve, have been known since about 1980. However, nearly all of them are either probabilistic, or they are inefficient in the following sense.

How does the running time of a computer algorithm grow as the input size increases? For primality testing, the input size is not the number concerned, but how many digits it has. The core distinction in such questions is between two classes of algorithms called P and not-P. If the running time grows like some fixed power of the input size, then the algorithm is class P; otherwise, it's not-P. Roughly speaking, class P algorithms are useful, whereas not-P algorithms are impractical, but there's a stretch of no-man's-land in between where other considerations come into play. Here P stands for 'polynomial time', a fancy way to talk about powers, and we return to the topic of efficient algorithms in chapter 11.

By the class P standard, trial division performs very badly. It's all right in the classroom, where the numbers that occur have two or three digits, but it's completely hopeless for 100-digit numbers. Trial division is firmly in the not-P class. In fact, the running time is roughly $10^{n/2}$ for an n-digit number, which grows faster than any fixed power of n. This type of growth, called exponential, is *really* bad, computational cloud-cuckoo-land.

Until the 1980s all known algorithms for primality testing, excluding probabilistic ones or those whose validity was unproved, had exponential growth rate. However, in 1983 an algorithm was found that lies tantalisingly in the no-man's-land adjacent to P territory: the aforementioned Adleman-Pomerance-Rumely test. An improved version by Henri Cohen and Hendrik Lenstra has running time n raised to the power log log n, where log denotes the logarithm. Technically, log log n can be as large as we wish, so this algorithm is not in class P. But that doesn't prevent it being practical: if n is a googolplex, 1 followed by 10^{100} zeros, then log log n is about 230. An old joke goes: 'It has been proved that log log n tends to infinity, but it has never been observed doing it.'

The first primality test in class P was discovered in 2002 by Manindra Agrawal and his students Neeraj Kayal and Nitin Saxena, who were undergraduates at the time. I've put some details in the Notes.[10] They proved that their algorithm has running time proportional to at most n^{12}; this was quickly improved to $n^{7.5}$. However, even though their algorithm is class P, hence classed as 'efficient', its advantages don't show up until the number n becomes very large indeed. It should beat the Adleman-Pomerance-Rumely test when the number of *digits* in n is about 10^{1000}. There isn't room to fit a number that big into a computer's memory, or, indeed, into the known universe. However, now that we *know* that a class P algorithm for primality testing exists, it becomes worthwhile to look for better ones. Lenstra and Pomerance reduced the power from 7.5 to 6. If various other conjectures about primes are true, then the power can be reduced to 3, which starts to look practical.

The most exciting aspect of the Agrawal-Kayal-Saxena algorithm, however, is not the result, but the method. It is simple – to mathematicians, anyway – and novel. The underlying idea is a variant of Fermat's theorem, but instead of working with numbers, Agrawal's team used a polynomial. This is a combination of powers of a variable x, such as $5x^3 + 4x - 1$. You can add, subtract, and multiply polynomials, and the usual algebraic laws remain valid. Chapter 3 explains polynomials in more detail.

This is a truly lovely idea: expand the domain of discourse and transport the problem into a new realm of thought. It is one of those ideas that are so simple you have to be a genius to spot them. It developed from a 1999 paper by Agrawal and his PhD supervisor Somenath Biswas, giving a probabilistic primality test based on an analogue of Fermat's theorem in the world of polynomials. Agrawal was convinced that the probabilistic element could be removed. In 2001 his students came up with a crucial, rather technical, observation. Pursuing that led the team into deep number-theoretic waters, but eventually everything was reduced to a single obstacle, the existence of a prime p such that $p - 1$ has a sufficiently large prime divisor. A bit of asking around and searching the Internet led to a theorem proved by Etienne Fouvry in 1985 using deep and technical methods. This was exactly what they needed to prove that their algorithm worked, and the final piece of the jigsaw slotted neatly into place.

In the days when number theory was safely tucked away inside its own little ivory tower, none of this would have mattered to the rest of the world. But over the last 20 years, prime numbers have become important in cryptography, the science of secret codes. Codes aren't just important for military use; commercial companies have secrets too. In this Internet age, we all do: we don't want criminals to gain access to our bank accounts, credit card numbers, or, with the growth of identity theft, the name of our cat. But the Internet is such a convenient way to pay bills, insure cars, and book holidays, that we have to accept some risk that our sensitive, private information might fall into the wrong hands.

Computer manufacturers and Internet service providers try to reduce that risk by making various encryption systems available. The involvement of computers has changed both cryptography and cryptanalysis, the dark art of code-breaking. Many novel codes have been devised, and one of the most famous, invented by Ron Rivest, Adi Shamir, and Leonard Adleman in 1978, uses prime numbers. Big ones, about a hundred digits long. The Rivest-Shamir-Adleman system is employed in many computer operating systems, is built into the main protocols for secure Internet communication, and is widely used by governments, corporations, and universities. That doesn't mean that every new result about primes is significant for the security of your Internet bank account, but it adds a definite frisson of excitement to any discovery that relates primes to computation. The Agrawal-Kayal-Saxena test is a case in point. Mathematically, it is elegant and important, but it has no direct practical significance.

It does, however, cast the general issue of Rivest-Shamir-Adleman cryptography in a new and slightly disturbing light. There is still no class P algorithm to solve Gauss's second problem, factorisation. Most experts think nothing of the kind exists, but they're not quite as sure as they used to be. Since new discoveries like the Agrawal-Kayal-Saxena test can lurk unsuspected in the wings, based on such simple ideas as polynomial versions of Fermat's theorem, cryptosystems based on prime factorisation might not be quite as secure as we fondly imagine. Don't reveal your cat's name on the Internet just yet.

Even the basic mathematics of primes quickly leads to more advanced concepts. The mystery becomes even deeper when we ask subtler questions. Euclid proved that the primes go on for ever, so we can't just list them all and stop. Neither can we give a simple, useful algebraic formula for successive primes, in the way that x^2 specifies squares. (There do exist simple formulas, but they 'cheat' by building the primes into the formula in disguise, and don't tell us anything new.[11]) To grasp the nature of these elusive, erratic numbers, we can carry out experiments, look for hints of structure, and try to prove that these apparent patterns persist no matter how large the primes become. For instance, we can ask how the primes are distributed among all whole numbers. Tables of primes strongly suggest that they tend to thin out as they get bigger. Table 1 shows how many primes there are in various ranges of 1000 consecutive numbers.

The numbers in the second column mostly decrease as we move down the rows, though sometimes there are brief periods when they go the other way: 114 is followed by 117, for instance. This is a symptom of the irregularity of the primes, but despite that, there is a clear general tendency for primes to become rarer as their size increases. The

range	number of primes
1–1000	168
1001–2000	135
2001–3000	127
3001–4000	119
4001–5000	118
5001–6000	114
6001–7000	117
7001–8000	106
8001–9000	110
9001–10,000	111

Table 1 The number of primes in successive intervals of 1000 numbers.

reason is not far to seek: the bigger a number becomes, the more potential factors there are. Primes have to avoid all of these factors. It's like fishing for non-primes with a net: the finer the net becomes, the fewer primes slip through.

The 'net' even has a name: the sieve of Eratosthenes. Eratosthenes of Cyrene was an ancient Greek mathematician who lived around 250 BC. He was also an athlete with interests in poetry, geography, astronomy, and music. He made the first reasonable estimate of the size of the Earth by observing the position of the Sun at noon in two different locations, Alexandria and Syene – present-day Aswan. At noon, the Sun was directly overhead at Syene, but about 7 degrees from the vertical at Alexandria. Since this angle is one fiftieth of a circle, the Earth's circumference must be 50 times the distance from Alexandria to Syene. Eratosthenes couldn't measure that distance directly, so he asked traders how long it took to make the journey by camel, and estimated how far a camel typically went in a day. He gave an explicit figure in a unit known as a *stadium*, but we don't know how long that unit was. Historians generally think that Eratosthenes's estimate was reasonably accurate.

His sieve is an algorithm to find all primes by successively eliminating all multiples of numbers already known to be prime. Figure 2 illustrates the method on the numbers up to 102, arranged to

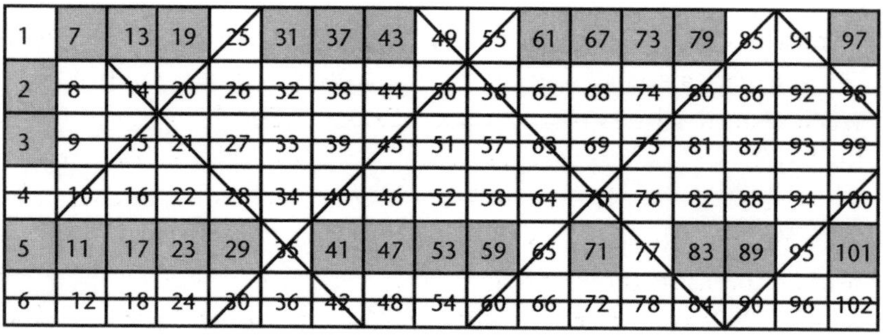

Fig 2 The sieve of Eratosthenes.

make the elimination process easy to follow. To see what's going on, I suggest you construct the diagram for yourself. Start with just the grid, omitting the lines that cross numbers out. Then you can add those lines one by one. Omit 1 because it's a unit. The next number is 2, so that's prime. Cross out all multiples of 2: these lie on the horizontal lines starting from 4, 6, and 8. The next number not crossed out is 3, so that's prime. Cross out all multiples of 3: these lie on the horizontal lines starting from 6, already crossed out, and 9. The next number not crossed out is 5, so that's prime. Cross out all multiples of 5: these lie on the diagonal lines sloping up and to the right, starting at 10. The next number not crossed out is 7, so that's prime. Cross out all multiples of 7: these lie on the diagonal lines sloping down and to the right, starting at 14. The next number not crossed out is 11, so that's prime. The first multiple of 11 that has not already been crossed out because it has a smaller divisor is 121, which is outside the picture, so stop. The remaining numbers, shaded, are the primes.

The sieve of Eratosthenes is not just a historical curiosity; it is still one of the most efficient methods known for making extensive lists of primes. And related methods have led to significant progress on what is probably the most famous unsolved great problem about primes: the Goldbach conjecture. The German amateur mathematician Christian Goldbach corresponded with many of the famous figures of his time. In 1742 he stated a number of curious conjectures about primes in a letter to Leonhard Euler. Historians later noticed that René Descartes had said much the same a few years before. The first of Goldbach's statements was: 'Every integer which can be written as the sum of two primes, can also be written as the sum of as many primes as one wishes, until all terms are units.' The second, added in the margin of his letter, was: 'Every integer greater than 2 can be written as the sum of three primes.' With today's definition of 'prime' there are obvious exceptions to these statements. For example, 4 is not the sum of three primes, because the smallest prime is 2, so the sum of three primes must be at least 6. But in Goldbach's day, the number 1 was considered to be prime. It is straightforward to rephrase his conjectures using the modern convention.

In his reply, Euler recalled a previous conversation with Goldbach, when Goldbach had pointed out that his first conjecture followed from a simpler one, his third conjecture: 'Every even integer is the sum of two primes.' With the prevailing convention that 1 is prime, this statement also implies the second conjecture, because any number can be written as either

$n + 1$ or $n + 2$ where n is even. If n is the sum of two primes, the original number is the sum of three primes. Euler's opinion of the third conjecture was unequivocal: 'I regard this as a completely certain theorem, although I cannot prove it.' That pretty much sums up its status today.

The modern convention, in which 1 is not prime, splits Goldbach's conjectures into two different ones. The even Goldbach conjecture states:

Every even integer greater than 2 is the sum of two primes.

The odd Goldbach conjecture is:

Every odd integer greater than 5 is the sum of three primes.

The even conjecture implies the odd one, but not conversely.[12] It is useful to consider both conjectures separately because we still don't know whether either of them is true. The odd conjecture seems to be slightly easier than the even one, in the sense that more progress has been made.

Some quick calculations verify the even Goldbach conjecture for small numbers:

$$4 = 2 + 2$$
$$6 = 3 + 3$$
$$8 = 5 + 3$$
$$10 = 7 + 3 = 5 + 5$$
$$12 = 7 + 5$$
$$14 = 11 + 3 = 7 + 7$$
$$16 = 13 + 3 = 11 + 5$$
$$18 = 13 + 5 = 11 + 7$$
$$20 = 17 + 3 = 13 + 7$$

It is easy to continue by hand up to, say, 1000 or so – more if you're persistent. For example $1000 = 3 + 997$, and $1{,}000{,}000 = 17 + 999{,}993$. In 1938 Nils Pipping verified the even Goldbach conjecture for all even numbers up to 100,000.

It also became apparent that as the number concerned gets bigger, there tend to be more and more ways to write it as a sum of primes. This makes sense. If you take a big even number, and keep subtracting primes in turn, how likely is it that *all* of the results will be composite? It takes just one prime to turn up among the resulting list of differences and the conjecture is verified for that number. Using statistical features of primes, we can assess the probability of such an outcome. The analysts Godfrey Harold Hardy and John Littlewood performed such a calculation in 1923, and derived a plausible but non-rigorous formula for the number of different ways to express a given even number n as a sum of two primes: approximately $n/[2(\log n)^2]$. This number increases as n becomes larger, and it also agrees with numerical evidence. But even if this calculation could be made rigorous, there might just be an occasional rare exception, so it doesn't greatly help.

The main obstacle to a proof of Goldbach's conjecture is that it combines two very different properties. Primes are defined in terms of multiplication, but the conjectures are about addition. So it is extraordinarily difficult to relate the desired conclusion to any reasonable features of primes. There seems to be nowhere to insert a lever. This must have been music to the ears of the publisher Faber & Faber in 2000, when it offered a million-dollar prize for a proof of the conjecture to promote the novel *Uncle Petros and Goldbach's Conjecture* by Apostolos Doxiadis. The deadline was tight: a solution had to be submitted before April 2002. No one made a successful claim to the prize, which is hardly surprising given that the problem has remained unsolved for over 250 years.

The Goldbach conjecture is often reformulated as a question about adding sets of integers together. The even Goldbach conjecture is the simplest example for this particular way of thinking, because we add just *two* sets of integers together. To do this, take any number from the first set, add any number from the second set, and then take the set of all such sums. For instance, the sum of {1, 2, 3} and {4, 5} contains $1 + 4, 2 + 4, 3 + 4, 1 + 5, 2 + 5, 3 + 5$, which is {5, 6, 7, 8}. Some numbers occur more than once, for instance $6 = 2 + 4 = 1 + 5$. I'll call this kind of repetition 'overlap'.

The even Goldbach conjecture can now be restated: if we add the set of primes to itself, the result contains every even number greater than 2. This reformulation may sound a bit trite – and is – but it moves the problem into an area where there are some powerful general theorems. The number 2 is a bit of a nuisance, but we can easily get rid of it. It is the only even prime, and if we add it to any other prime the result is odd. So as far as the even Goldbach conjecture is concerned, we can forget about 2. However, we need $2 + 2$ to represent 4, so we must also restrict attention to even numbers that are at least 6.

As a simple experiment, consider the even numbers up to and including 30. There are nine odd primes in this range: {3, 5, 7, 11, 13, 17, 19, 23, 29}. Adding them gives Figure 3: I've marked the sums that are less than or equal to 30 (a range of even numbers that includes all primes up to 29) in bold. Two simple patterns appear. The whole table is symmetric about its main diagonal because $a + b = b + a$. The bold numbers occupy roughly the top left half of the table, above the thick (diagonal) line. If anything, they tend to bulge out beyond it in the middle. This happens because on the whole, large primes are rarer than small ones. The extra region of the bulge more than compensates for the two 32s at top right and bottom left.

Now we make some rough estimates. I could be more precise, but these are good enough. The number of slots in the table is $9 \times 9 = 81$.

	3	5	7	11	13	17	19	23	29
3	6	8	10	14	16	20	22	26	32
5	8	10	12	16	18	22	24	28	34
7	10	12	14	18	20	24	26	30	36
11	14	16	18	22	24	28	30	34	40
13	16	18	20	24	26	30	32	36	42
17	20	22	24	28	30	34	36	40	46
19	22	24	26	30	32	36	38	42	48
23	26	28	30	34	36	40	42	46	52
29	32	34	36	40	42	46	48	52	58

Fig 3 Sums of pairs of primes up to 30. Boldface: sums that are 30 or smaller. Thick line: diagonal. Shaded region: eliminating symmetrically related pairs. The shaded region is slightly more than one quarter of the square.

About half of the numbers in those slots are in the top left triangle. Because of the symmetry, these arise in pairs except along the diagonal, so the number of unrelated slots is about 81/4, roughly 20. The number of even integers in the range from 6 to 30 is 13. So the 20 (and more) boldface sums have to hit only 13 even numbers. There are more potential sums of two primes in the right range than there are even numbers. It's like throwing 20 balls at 13 coconuts at the fair. You have a reasonable chance of hitting a lot of them. Even so, you could miss a few coconuts. Some even numbers might still be missing.

In this case they're not, but this kind of counting argument can't eliminate that possibility. However, it does tell us that there must be quite a bit of overlap, where the same boldface number occurs several times in the relevant quarter of the table. Why? Because 20 sums have to fit into a set with only 13 members. So on average each boldface number appears about 1.5 times. (The actual number of sums is 27, so a better estimate shows that each boldface number appears twice.) If any even numbers are missing, the overlap must be bigger still.

We can play the same game with a larger upper limit – say 1

million. A formula called the prime number theorem, chapter 9, provides a simple estimate for the number of primes up to any given size x. The formula is $x/\log x$. Here, the estimate is about 72,380. (The exact figure is 78,497.) The corresponding shaded region occupies about one quarter of the table, so it provides about $n^2/4 = 250$ billion boldface numbers: sums of two primes in this range. This is vastly larger than the number of even numbers in the range, which is half a million. Now the amount of overlap has to be gigantic, with each sum occurring on average 500,000 times. So the chance of any particular even number escaping is greatly reduced.

With more effort, we can turn this approach into an estimate of the probability that some even number in a given range is not the sum of two primes, assuming that the primes are distributed at random and with frequencies given by the prime number theorem – that is, about $x/\log x$ primes less than any given x. This is what Hardy and Littlewood did. They knew that their approach wasn't rigorous, because primes are defined by a specific process and they're not actually random. Nevertheless, it's sensible to expect the actual results to be consistent with this probabilistic model, because the defining property of primes seems to have very little connection with what happens when we add two of them together.

Several standard methods in this area adopt a similar point of view, but taking extra care to make the argument rigorous. Sieve methods, which build on the sieve of Eratosthenes, are examples. General theorems about the density of numbers in sums of two sets – the proportion of numbers that occur, as the sets become very large – provide other useful tools.

When a mathematical conjecture eventually turns out to be correct, its history often follows a standard pattern. Over a period of time, various people prove the conjecture to be true provided special restrictions apply. Each such result improves on the previous one by relaxing some restrictions, but eventually this process runs out of steam. Finally, a new and much cleverer idea completes the proof.

For example, a conjecture in number theory may state that every positive integer can be represented in some manner using, say, six special numbers (prime, square, cube, whatever). Here the key features are *every* positive integer and *six* special numbers. Initial advances lead to much weaker results, but successive stages in the process slowly improve them.

The first step is often a proof along these lines: every positive integer that is not divisible by 3 or 11, except for some finite number of them, can be represented in terms of some gigantic number of special numbers – say 10^{666}. The theorem typically does not specify how many exceptions there are, so the result cannot be applied directly to any specific integer. The next step is to make the bound effective: that is, to prove that every integer greater than $10^{10^{42}}$ can be so represented. Then the restriction on divisibility by 3 is eliminated, followed by a similar advance for 11. After that, successive authors reduce one of the numbers 10^{666} or $10^{10^{42}}$, often both. A typical improvement might be that every integer greater than 5.8×10^{17} can be represented using at most 4298 special numbers, for instance.

Meanwhile, other researchers are working upwards from small numbers, often with computer assistance, proving that, say, every number less than or equal to 10^{12} can be represented using at most six special numbers. Within a year, 10^{12} has been improved in five stages, by different researchers or groups, to 11.0337×10^{29}. These improvements are neither routine nor easy, but the way they are achieved involves intricate special methods that provide no hint of a more general approach, and each successive contribution is more complicated and longer. After a few years of this kind of incremental improvement, applying the same general ideas but with more powerful computers and new tweaks, this number has risen to 10^{43}. But now the method grinds to a halt, and everyone agrees that however much tweaking is done, it will never lead to the full conjecture.

At that point the conjecture disappears from view, because no one is working on it any more. Sometimes, progress pretty much stops. Sometimes, twenty years pass with nothing new ... and then, apparently from nowhere, Cheesberger and Fries announce that by reformulating the conjecture in terms of complex meta-ergodic quasiheaps and applying byzantine quisling theory, they have obtained a complete proof. After several years arguing about fine points of logic, and plugging a few gaps, the mathematical community accepts that the proof is correct, and immediately asks if there's a better way to achieve the same result, or to push it further.

You will see this pattern work itself out many times in later chapters. Because such accounts become tedious, no matter how proud Buggins and Krumm are of their latest improvement of the exponent in the Jekyll-Hyde conjecture from 1.773 to $1.771 + \varepsilon$ for any positive ε, I will describe a few representative contributions and leave out the rest. This is not to deny the importance of the work of Buggins and Krumm. It may even have paved the way to the great Cheesberger-Fries breakthrough. But only experts, following the developing story, are likely to await the next tiny improvement with bated breath.

In future I'll provide less detail, but let's see how it goes for Goldbach.

Theorems that go some way towards establishing Goldbach's conjecture have been proved. The first big breakthrough came in 1923, when Hardy and Littlewood used their analytic techniques to prove the odd Goldbach conjecture for all sufficiently large odd numbers. However, their proof relied on another big conjecture, the generalised Riemann hypothesis, which we discuss in chapter 9. This problem is still open, so their approach had a significant gap. In 1930 Lev Schnirelmann bridged the gap using a fancy version of their reasoning, based on sieve methods. He proved that a nonzero proportion of all numbers can be represented as a sum of two primes. By combining this result with some generalities about adding sequences together, he proved that there is some number C such that every integer greater than 1 is a sum of at most C prime numbers. This number became known as Schnirelmann's constant. Ivan Matveyevich Vinogradov obtained similar results in 1937, but his method also did not specify how big 'significantly large' is. In 1939 K. Borozdin proved that it is no greater than $3^{14,348,907}$. By 2002 Liu Ming-Chit and Wang Tian-Ze had reduced this 'upper bound' to e^{3100}, which is about 2×10^{1346}. This is a lot smaller, but it is still too big for the intermediate numbers to be checked by computer.

In 1969 N.I. Klimov obtained the first specific estimate for Schnirelmann's constant: it is at most 6 billion. Other mathematicians reduced that number considerably, and by 1982 Hans Riesel and Robert Vaughan had brought it down to 19. Although 19 is a lot better than 6 billion, the evidence pointed to

Schnirelmann's constant being a mere 3. In 1995 Leszek Kaniecki reduced the upper bound to 6, with five primes for any odd number, but he had to assume the truth of the Riemann hypothesis. His results, combined with J. Richstein's numerical verification of the Riemann hypothesis up to 4×10^{14}, would prove that Schnirelmann's constant is at most 4, again assuming the Riemann hypothesis. In 1997 Jean-Marc Deshouillers, Gove Effinger, Herman te Riele, and Dmitrii Zinoviev showed that the generalised Riemann hypothesis (chapter 9) implies the odd Goldbach conjecture. That is, every odd number except 1, 3, and 5 is the sum of three primes.

Since the Riemann hypothesis is currently not proved, it is worth trying to remove this assumption. In 1995 the French mathematician Olivier Ramaré reduced the upper estimate for representing odd numbers to 7, without using the Riemann hypothesis. In fact, he proved something stronger: every even number is a sum of at most six primes. (To deal with odd numbers, subtract 3: the result is even, so it is a sum of six or fewer primes. The original number is this sum plus the prime 3, requiring seven or fewer primes.) The main breakthrough was to improve existing estimates for the proportion of numbers, in some specified range, that are the sum of two primes. Ramaré's key result is that for any number n greater than e^{67} (about 1.25×10^{29}), at least one fifth of the numbers between n and $2n$ are the sum of two primes. Using sieve methods, in conjunction with a theorem of Hans-Heinrich Ostmann about sums of sequences, refined by Deshouillers, this leads to a proof that every even number greater than 10^{30} is a sum of at most six primes.

The remaining obstacle is to deal with the gap between 4×10^{14}, where Jörg Richstein had checked the theorem by computer, and 10^{30}. As is common, the numbers are too big for a direct computer search, so Ramaré proved a series of specialised theorems about the number of primes in small intervals. These theorems depend on the truth of the Riemann hypothesis up to specific limits, which can be verified by computer. So the proof consists mainly of conceptual pencil-and-paper deductions, with computer assistance in this particular respect. Ramaré ended his paper by pointing out that in principle a similar approach could reduce the number of primes from 7 to 5. However, there were huge practical obstacles, and he wrote that such a proof 'can not be reached by today's computers'.

In 2012 Terence Tao overcame those difficulties with some new and very different ideas. He posted a paper on the Internet, which as I write is under review for publication. Its main theorem is: every odd number is a sum of at most five primes. This reduces Schnirelmann's constant to 6. Tao is renowned for his ability to solve difficult problems in many areas of mathematics. His proof throws several powerful techniques at the problem, and requires computer assistance. If the number 5 in Tao's theorem could be reduced to 3, the odd Goldbach conjecture would be proved, and the bound on Schnirelmann's constant reduced to 4. Tao suspects that it should be possible to do this, although further new ideas will be needed.

The even Goldbach conjecture seems harder still. In 1998 Deshouillers, Saouter, and te Riele verified it for all even numbers up to 10^{14}. By 2007, Tomás Oliveira e Silva had improved that to 10^{18}, and his computations continue. We know that every even integer is the sum of at most six primes – proved by Ramaré in 1995. In 1973 Chen Jing-Run proved that every sufficiently large even integer is the sum of a prime and a semiprime (either a prime or a product of two primes). Close, but no cigar. Tao has stated that the even Goldbach conjecture is beyond the reach of his methods. Adding three primes together creates far more overlap in the resulting numbers – in the sense discussed in connection with Figure 3 – than the two primes needed for the even Goldbach conjecture, and Tao's and Ramaré's methods exploit this feature repeatedly.

In a few years' time, then, we may have a complete proof of the odd Goldbach conjecture, in particular implying that every even number is the sum of at most four primes. But the even Goldbach conjecture will probably still be just as baffling as it was for Euler and Goldbach.

In the 2300 years since Euclid proved several basic theorems about primes, we have learned a great deal more about these elusive, yet vitally important, numbers. But what we now know puts into stark perspective the long list of what we don't know.

We know, for instance, that there are infinitely many primes of the form $4k + 1$ and $4k + 3$; more generally, that any arithmetic sequence[13] $ak + b$ for fixed a and b contains infinitely many primes provided a and b have no common factor. For instance, suppose that $a = 18$. Then

$b = 1, 5, 7, 11, 13$, or 17. Therefore there exist infinitely many primes of each of the forms $18k + 1$, $18k + 5$, $18k + 7$, $18k + 11$, $18k + 13$, or $18k + 17$. This is not true for, say, $18k + 6$, because this is a multiple of 6. No arithmetic sequence can contain *only* primes, but a recent major breakthrough, the Green-Tao theorem, shows that the set of primes contains arbitrarily long arithmetic sequences. The proof, obtained in 2004 by Ben Green and Tao, is deep and difficult. It gives us hope: difficult open questions, however impenetrable they may appear, can sometimes be answered.

Putting on our algebraist's hat we immediately wonder about more complicated formulas involving k. There are no primes of the form k^2, and none except 3 for the form $k^2 - 1$, because these expressions factorise. However, the expression $k^2 + 1$ does not have obvious factors, and here we can find plenty of primes:

$$2 = 1^2 + 15 = 2^2 + 117 = 4^2 + 137 = 6^2 + 1$$

and so on. A larger example of no special significance is

$$18,672,907,718,657 = (4,321,216)^2 + 1$$

It is conjectured that infinitely many such primes exist, but no such statement has yet been proved for any specific polynomial in which k occurs to a higher power than the first. A very plausible conjecture is the one made by V. Bouniakowsky in 1857: any polynomial in k that does not have obvious divisors represents infinitely many primes. The exceptions here include not only reducible polynomials, but ones like $k^2 + k + 2$ which is always divisible by 2, despite having no algebraic factors.

Some polynomials seem to have special properties. The classic case is $k^2 + k + 41$, which is prime for $k = 0, 1, 2, ..., 40$, and indeed also for $k = -1, -2, ..., -40$. Long runs of primes for consecutive values of k are rare, and a certain amount is known about them. But the whole area is very mysterious.

Almost as famous as the Goldbach conjecture, and apparently just as hard, is the twin primes conjecture: there are infinitely many pairs of primes that differ by 2. Examples are

$$3, 5 \qquad 5, 7 \qquad 11, 13 \qquad 17, 19$$

The largest known twin primes (as of January 2012) are

$$3,756,801,695,685 \times 2^{666,669} \pm 1$$

which have 200,700 decimal digits. They were found by the PrimeGrid distributed computing project in 2011. In 1915, Viggo Brun used a variant of the sieve of Eratosthenes to prove that the sum of reciprocals of all twin primes converges, unlike the sum of the reciprocals of all primes. So in this sense, twin primes are relatively rare. He also proved, using similar methods, that there exist infinitely many integers n such that n and $n+2$ have at most nine prime factors. Hardy and Littlewood used their heuristic methods to argue that the number of twin prime pairs less than x should be asymptotic to

$$2a \frac{n}{(\log n)^2}$$

where a is a constant whose value is about 0.660161. The underlying idea is that for this purpose primes can be assumed to arise at random, at a rate that makes the number of primes up to x approximately equal to $x/\log x$. There are many similar conjectures and heuristic formulas, but again, no rigorous proofs.

Indeed, there are hundreds of open questions about primes. Some are just curios, some are deep and significant. We will meet some of the latter in chapter 9. Despite all of the advances mathematicians have made over the last two and a half millennia, the humble primes have lost none of their allure and none of their mystery.

3

The puzzle of pi
Squaring the Circle

PRIMES ARE AN OLD IDEA, but circles are even older. Circles led to a great problem that took more than 2000 years to solve. It is one of several related geometric problems that have come down to us from antiquity. The central character in the story is the number π (Greek 'pi') which we meet at school in connection with circles and spheres. Numerically it is 3.14159 and a bit; often the approximation 22/7 is used. The digits of π never stop, and they never repeat the same sequence over and over again. The current record for calculating digits of π is 10 trillion digits, by Alexander Yee and Chigeru Kondo in October 2011.[14] Computations like this are significant as ways to test fast computers, or to inspire and test new methods to calculate π, but very little hinges on the numerical results. The reason for being interested in π is not to calculate the circumference of a circle. The same strange number appears all over mathematics, not just in formulas related to circles and spheres, and it leads into very deep waters indeed. The school formulas are important, even so, and they reflect π's origins in Greek geometry.

There, one of the great problems was the unsolved task of squaring the circle. This phrase is often employed colloquially to indicate a wrong-headed approach to something, rather like trying to fit a square peg into a round hole. Like many common phrases extracted from science, this one's meaning has changed over the centuries.[15] In Greek times, trying to square the circle was a perfectly reasonable idea. The difference in the two shapes – straight or curved – is totally irrelevant:

similar problems have valid solutions.[16] However, it eventually turned out that this particular problem cannot be solved using the specified methods. The proof is ingenious and technical, but its general nature is comprehensible.

In mathematics, squaring the circle means constructing a square whose *area* is the same as that of a given circle, using the traditional methods of Euclid. Greek geometry actually permitted other methods, so one aspect of the problem is to pin down which methods are to be used. The impossibility of solving the problem is then a statement about the limitations of those methods; it doesn't imply that we can't work out the area of a circle. We just have to find another approach. The impossibility proof explains why the Greek geometers and their successors failed to find a construction of the required kind: there isn't one. In retrospect, that explains why they had to introduce more esoteric methods. So the solution, despite being negative, clears up what would otherwise be a big historical puzzle. It also stops people wasting time in a continuing search for a construction that doesn't exist – except for a few hardy souls who regrettably seem unable to get the message, no matter how carefully it is explained.[17]

In Euclid's *Elements* the traditional methods for constructing geometric figures are idealised versions of two mathematical instruments: the ruler and the compass. To be pedantic, compass*es*, for the same reason that you cut paper with scissor*s*, not with a scissor – but I will follow common parlance and avoid the plural. These instruments are used to 'draw' diagrams on a notional sheet of paper, the Euclidean plane.

Their form determines what they can draw. A compass comprises two rigid rods, hinged together. One has a sharp point, the other holds a sharp pencil. The instrument is used to draw a circle, or part of one, with a specific centre and a specific radius. A ruler is simpler: it has a straight edge, and is used to draw a straight line. Unlike the rulers you buy in stationery shops, Euclid's rulers have no marks on them, and this is an important restriction for the mathematical analysis of what they can create.

The sense in which the geometer's ruler and compass are idealisations is straightforward: they are assumed to draw infinitely thin lines. Moreover, the straight lines are exactly straight and the circles are perfectly round. The paper is perfectly flat and even. The other key ingredient of Euclid's geometry is the notion of a point, another ideal. A point is a dot on the paper, but it is a physical impossibility: it has no size. 'A point', said Euclid, in the first sentence of the *Elements*, 'is that which has no part.' This sounds a bit like an atom, or if you're clued into modern physics, a subatomic particle, but compared to a geometric point, those are gigantic. From an everyday human perspective, however, Euclid's ideal point, an atom, and a pencil dot on a sheet of paper, are similar enough for the purposes of geometry.

These ideals are not attainable in the real world, however carefully you make the instruments and sharpen the pencil, and however smooth you make the paper. But idealism can be a virtue, because these requirements make the mathematics much simpler. For instance, two pencil lines cross in a small fuzzy region shaped like a parallelogram, but mathematical lines meet at a single point. Insights gained from ideal circles and lines can often be transferred to real, imperfect ones. This is how mathematics works its magic.

Two points determine a (straight) line, the unique line that passes through them. To construct the line, place your ideal ruler so that it passes through the two points, and run your ideal pencil along it. Two points also determine a circle: choose one as the centre, and place the compass point there; then adjust it so that the tip of the pencil lies on the other point. Now swing the pencil round in an arc, keeping the central point fixed. Two lines determine a unique point, where they cross, unless they are parallel, in which case they don't cross, but a Pandora's box of logical issues yawns wide. A line and a circle determine two points, if they cross; one point, if the line cuts the circle at a tangent; nothing at all if the circle is too small to meet the line. Similarly two circles either meet in two points, one, or none.

Distance is a fundamental concept in the modern treatment of Euclidean geometry. The distance between any two points is measured along the line that joins them. Euclid managed to get his geometry working without an explicit concept of distance, by finding a way to say that two line segments have the *same* length without defining length itself. In fact, this is easy: just stretch a compass between the ends of one segment, transfer it to the second, and see if the ends fit. If they do, the lengths are equal; if they don't, they're not. At no stage do you measure an actual length.

From these basic ingredients, geometers can build up more interesting shapes and configurations. Three points determine a triangle unless they all lie on the same line. When two lines cross, they form an angle. A right angle is especially significant; a straight line corresponds to two right angles joined together. And so on, and so on, and so on. Euclid's *Elements* consists of 13 books, delving ever deeper into the consequences of these simple beginnings.

The bulk of the *Elements* consists of theorems – valid features of geometry. But Euclid also explains how to solve geometric problems, using 'constructions' based on ruler and compass. Given two points joined by a segment of a line, construct their midpoint. Or trisect the segment: construct a point exactly one third of the way along it. Given an angle, construct one that bisects it – is half the size. But some simple constructions proved elusive. Given an angle, construct one that trisects it – is one third the size. You can do that for line segments, but no one could find a method for angles. Approximations, as close as you wish, yes. Exact constructions using only an unmarked ruler and a compass: no. However, no one really needs to trisect angles exactly anyway, so this particular issue didn't cause much trouble.

More embarrassing was a construction that could not be ignored: given a circle, construct a square that has the same area. This is the problem of squaring the circle. From the Greek point of view, if you couldn't solve that, you weren't entitled to claim that a circle *had* an area. Even though it visibly encloses a well-defined space, and intuitively the area is *how much* space. Euclid and his successors, notably Archimedes, settled for a pragmatic solution: assume circles have areas, but don't expect to be able to construct squares with the same area. You can still say a lot; for instance, you can prove, in full logical rigour, that the area of a circle is proportional to the square of its diameter. What you can't do, without squaring the circle, is to construct a line whose length is the constant of proportionality.

The Greeks couldn't square the circle using ruler and compass, so they settled for other methods. One used a curve called a quadratrix.[18] The importance they attached to using only ruler and compass was exaggerated by some later commentators, and it's not even clear that the Greeks considered squaring the circle to be a vital issue. By the nineteenth century, however, the problem was becoming a major nuisance. Mathematics that was unable to answer such a straightforward question was like a cordon bleu cook who didn't know how to boil an egg.

Squaring the circle sounds like a problem in geometry. That's because it is a problem in geometry. But its solution turned out to lie not in geometry at all, but in algebra. Making unexpected connections between apparently unrelated areas of mathematics often lies at the heart of solving a great problem. Here, the connection was not entirely unprecedented, but its link to squaring the circle was not at first appreciated. Even when it was, there was a technical difficulty, and dealing with that required yet another area of mathematics: analysis, the rigorous version of calculus. Ironically, the first breakthrough came from a fourth area: number theory. And it solved a geometric problem that the Greeks would never in their wildest dreams have believed to possess a solution, and as far as we can tell never thought about: how to construct, with ruler and compass, a regular polygon with 17 sides.

It sounds mad, especially if I add that no such construction exists for regular polygons with 7, 9, 11, 13, or 14 sides, but one does for 3, 4, 5, 6, 8, 10, and 12. However, there is method behind the madness, and it is the method that enriched mathematics.

First: what is a regular polygon? A polygon is a shape bounded by straight lines. It is regular if those lines have equal length and meet at equal angles. The most familiar example is the square: all four sides have the same length and all four angles are right angles. There are other shapes with four equal sides or four equal angles: the rhombus and rectangle respectively. Only a square has both features. A regular 3-sided polygon is an equilateral triangle, a regular 5-sided polygon is a regular pentagon, and so on, Figure 4. Euclid provides ruler-and-compass constructions for regular polygons with 3, 4, and 5 sides. The Greeks also knew how to repeatedly double the number of sides, giving 6, 8, 10, 12, 16, 20, and so on. By combining the constructions for 3- and 5-sided regular polygons they could obtain a 15-sided one. But there, their knowledge stopped. And for about 2000 years that's how it remained. No one imagined that any other numbers were feasible.

They didn't even ask, it just seemed obvious that nothing more could be done.

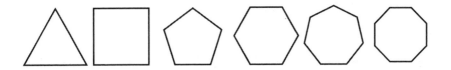

Fig 4 The first few regular polygons. *From left to right*: equilateral triangle, square, pentagon, hexagon, heptagon, octagon.

It took one of the greatest mathematicians who have ever lived to think the unthinkable, ask the unaskable, and discover a truly astonishing answer. Namely, Gauss Carl Friedrich. Gauss was born into a poor, working-class family in the city of Braunschweig (Brunswick) in Germany. His mother Dorothea could not read or write, and failed to write down the date of his birth, but she did remember that it was on a Wednesday, eight days before the feast of the ascension, in 1777. Gauss later worked out the exact date from a mathematical formula he devised for the date of Easter. His father Gebhard came from a farming family, but made a living in a series of low-level jobs: gardener, canal labourer, street butcher, funeral parlour accountant. Their son was a child prodigy who is reputed to have corrected his father's arithmetic at the age of three, and his abilities, which extended to languages as well as mathematics, led the Duke of Braunschweig to fund his university studies at the Collegium Carolinum. While an undergraduate Gauss independently rediscovered several important mathematical theorems that had been proved by illustrious people such as Euler. But his theorem about the regular 17-sided polygon came as a bolt from the blue.

By then, the close link between geometry and algebra had been understood for 140 years. In an appendix to *Discours de la Méthode* ('Discourse on the method') René Descartes formalised an idea that had been floating around in rudimentary form for some time: the notion of a coordinate system. In effect, this takes Euclid's barren plane, a blank sheet of paper, and turns it into paper ruled into squares, which engineers and scientists call graph paper. Draw two straight lines on the paper, one horizontal, the other vertical: these are

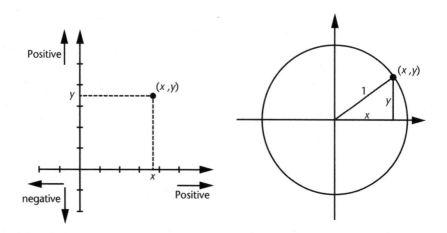

Fig 5 *Left*: Coordinates in the plane. *Right*: How to derive the equation for the unit circle.

called axes. Now you can pin down the location of any point of the plane by asking how far it lies in the direction along the horizontal axis, and how far up the vertical axis, Figure 5 (left). These two numbers, which may be positive or negative, provide a complete description of the point, and they are called its coordinates.

All geometric properties of points, lines, circles, and so on can be translated into algebraic statements about the corresponding coordinates. It's very difficult to talk meaningfully about these connections without using some actual algebra – just as it's hard to talk sensibly about football without mentioning the word 'goal'. So the next few pages will include some formulas. They are there to ensure that the main players in the drama have names and the relationship between them is clear. 'Romeo' is so much simpler to follow than 'the son of an Italian patriarch who falls in love with his father's sworn enemy's beautiful daughter'. Our Romeo will bear the prosaic name x, and his Juliet will be y.

As an example of how geometry converts into algebra, Figure 5 (right) shows how to find the equation for a circle of unit radius centred at the origin, where the two axes cross. The marked point has coordinates (x, y), so the right-angled triangle in the figure has horizontal side of length x and vertical side of length y. The longest side of the triangle is the radius of the circle, which is 1. Pythagoras's theorem now tells us that the sum of the squares of the two coordinates is 1. In symbols, a point with coordinates x and y lies on the circle if (and only if) it satisfies the condition $x^2 + y^2 = 1$. This symbolic characterisation of the circle is brief and precise, and it shows that we really are talking algebra. Conversely, any algebraic property of pairs of numbers, any equation involving x and y, can be reinterpreted as a geometric statement about points, lines, circles, or more elaborate curves.[19]

The basic equations of algebra involve polynomials, combinations of powers of an unknown quantity x, where each power is multiplied by some number, called a coefficient. The largest power of x that occurs is the degree of the polynomial. For example, the equation

$$x^4 - 3x^3 - 3x^2 + 15x - 10 = 0$$

involves a polynomial starting with x^4, so its degree is 4. The coefficients are 1, -3, -3, 15, and -10. There are four distinct solutions: $x = 1, 2, \sqrt{5}$, and $\sqrt{5}$. For these numbers the left-hand side of the equation is equal to zero – the right-hand side. Polynomials of degree 1, like $7x + 2$, are said to be linear, and they involve only the first power of the unknown. Equations of degree 2, like $x^2 - 3x + 2$, are said to be quadratic, and they involve the second power – the square. The equation for a circle involves a second variable, y. However, if we know a second equation relating x and y, for example the equation defining some straight line, then we can solve for y in terms of x and reduce the equation for a circle to one that involves only x. This new equation tells us where the line meets the circle. In this case the new equation is quadratic, with two solutions; this is how the algebra reflects the geometry, in which a line meets the circle at two distinct points.

This feature of the algebra has an important implication for ruler-and-compass constructions. Such a construction, however complicated, breaks up into a sequence of simple steps. Each step produces new points at places where two lines, two circles, or a line and a circle, meet. Those lines and circles are determined by previously constructed points. By translating geometry into algebra, it can be proved that the algebraic equation that corresponds to the intersection of two lines is always linear, while that for a line and a circle, or two circles, is quadratic. Ultimately this happens because the equation for a circle involves x^2 but no higher power of x. So every individual step in a construction corresponds to solving an equation of degree 1 or 2 only.

More complex constructions are sequences of these basic operations, and a certain amount of algebraic technique lets us deduce that each coordinate of any point that can be constructed by ruler and compass is a solution of a polynomial equation, with integer coefficients, whose degree is a power of 2. That is, the degree has to be one of the numbers 1, 2, 4, 8, 16, and so on.[20] This condition is necessary for a construction to exist, but it can be beefed up into a precise characterisation of which regular polygons are constructible. Suddenly a tidy algebraic condition emerges from a complicated geometric muddle – and it applies to *any construction whatsoever*. You don't even need to know what the construction is: just that it uses only ruler and compass.

Gauss was aware of this elegant idea. He also knew (indeed, any competent mathematician would quickly realise) that the question of which regular polygons can be constructed by ruler and compass boils down to a special case, when the polygon has a prime number of sides. To see why, think of a composite number like 15, which is 3×5. Any hypothetical construction of a 15-sided regular polygon automatically yields a 3-sided one (consider every fifth vertex) and a 5-sided one (consider every third vertex), Figure 6. With a bit more effort you can combine constructions for a 3-gon and a 5-gon to get a 15-gon.[21] The numbers 3 and 5 are prime, and the same idea applies in general. So Gauss focused on polygons with a prime number of sides, and asked what the relevant equation looked like. The answer was surprisingly neat. Constructing a regular 5-sided polygon, for example, is equivalent to solving the equation $x^5 - 1 = 0$. Replace 5 by any other prime, and the corresponding statement is true.

The degree of this polynomial is 5, which is *not* one of the powers of 2 that I listed; even so, a construction exists. Gauss quickly figured out why: the equation splits into two pieces, one of degree 1 and the other of degree 4. Both 1 and 4 are powers of 2, and it turns out that the degree-4 equation is the crucial one. To see why, we need to connect the equation to the geometry. That involves a new kind of number, one that is largely ignored in school mathematics but is indispensable for

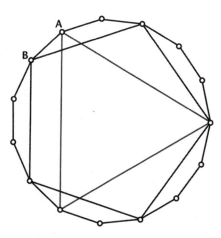

Fig 6 Constructing an equilateral triangle and a regular pentagon from a regular 15-gon. For the reverse, observe that A and B are consecutive points on the regular 15-gon.

anything beyond that. They are called complex numbers, and their defining feature is that in the complex number system −1 has a square root.[22]

An ordinary 'real' number is either positive or negative, and either way, its square is positive, so −1 can't be the square of any real number. This is such a nuisance that mathematicians invented a new kind of 'imaginary' number whose square is −1. They needed a new symbol for it, so they called it i (for 'imaginary'). The usual operations of algebra − adding, subtracting, multiplying, dividing − lead to combinations of real and imaginary numbers such as 3 + 2i. These are said to be complex, which doesn't mean 'complicated', but indicates that they come in two parts: 3 and 2i. Real numbers lie on the famous number line, like the numbers on a ruler. Complex numbers lie in a number plane, in which an imaginary ruler is placed at right angles to a real one, and the two together form a system of coordinates, Figure 7 (left).

For the last 200 years, mathematicians have considered complex numbers to be fundamental to their subject. We now recognise that logically they are on the same footing as the more familiar 'real' numbers − which, like all mathematical structures, are abstract concepts, not real physical things. Complex numbers were in widespread use before the time of Gauss, but their status was still

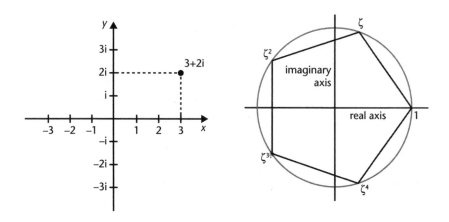

Fig 7 *Left:* The complex plane. *Right:* The complex fifth roots of unity.

mysterious until Gauss and several others demystified them. The source of their attraction was paradoxical: despite the mystery surrounding their meaning, complex numbers were much better behaved than real numbers. They supplied a missing ingredient that the real numbers lacked. They provided a complete set of solutions for an algebraic equation.

Quadratic equations are the simplest example. Some quadratics have two real solutions, while others have none. For example $x^2 - 1 = 0$ has the solutions 1 and -1, but $x^2 + 1 = 0$ has no solutions. In between is $x^2 = 0$, whose sole solution is 0, but there is a sense in which this is the same solution 'repeated twice'.[23] If we allow complex solutions, however, then $x^2 + 1 = 0$ also has two solutions: i and $-$i. Gauss had no qualms about using complex numbers; in fact, his doctoral thesis provided the first logically sound proof of the fundamental theorem of algebra: the number of complex solutions to any polynomial equation (with multiplicities counted correctly) is equal to the degree of the equation. So quadratics (degree 2) always have two complex solutions, cubics (degree 3) always have three complex solutions, and so on.

The equation $x^5 - 1 = 0$, which I claimed defines a regular pentagon, has degree 5. Therefore it has five complex solutions. There is just one real solution: $x = 1$. What about the other four? They provide four vertexes of a perfect regular pentagon in the complex plane, with $x = 1$ being the fifth, Figure 7 (right). This correspondence is an example of mathematical beauty: an elegant geometric shape becomes an elegant equation.

Now, the equation whose solutions are these five points has degree 5, which is not a power of 2. But, as mentioned earlier, the degree-5 equation splits into two pieces with degrees 1 and 4, called its irreducible factors:

$$x^5 - 1 = (x - 1)(x^4 + x^3 + x^2 + x + 1)$$

('Irreducible' means that no further factors exist, just like prime numbers.) The first factor yields the real solution $x = 1$. The other factor yields the four complex solutions and the other four vertexes of the pentagon. So everything makes much more sense, and is far more elegant, when we use complex numbers.

It is often difficult to reconstruct how the mathematicians of the past arrived at new discoveries, because they had a habit of presenting only the final outcome of their deliberations, not the many false steps they took along the way. This problem is often compounded, because the natural thought patterns in past ages were different from today's. Gauss in particular was notorious for covering his tracks and publishing only his final, highly polished, analysis. But when it comes to Gauss's research on the 17-sided polygon, we are on fairly safe ground; the analysis that he eventually published provides several useful clues.

His starting-point was not new. Several earlier mathematicians were well aware that the above analysis of regular pentagons works in complete generality. Constructing a polygon with any number n of sides is equivalent to solving the equation $x^n - 1 = 0$ in complex numbers. Moreover, this polynomial factorises as

$$(x - 1)(x^{n-1} + x^{n-2} + \cdots + x^2 + x + 1)$$

Again the first factor gives the real solution $x = 1$ and the remaining $n - 1$ solutions come from the second factor. When n is odd, these are all complex; when n is even, one of them is a second real solution $x = -1$.

What Gauss noticed, and everyone else had missed, is that

sometimes the second factor can be expressed using a series of quadratic equations. Not by representing it as a product of simpler factors, because that's not possible, but by using equations whose coefficients solve other equations. The key fact here – the weak point in the problem – is an elegant property of algebraic equations, which arises when we solve several of them in turn in this manner. The calculation is always equivalent to solving a single equation, but the degree generally gets bigger. So the price we pay for having fewer equations is an increase in the degree. It can be messy, but there is one feature that we can predict: how big the degree becomes. Just multiply together the degrees of the successive polynomials.

If they are all quadratics, the result is $2 \times 2 \times \cdots \times 2$, a power of 2. So $n - 1$ must be a power of 2 if a construction exists. However, this condition is not always sufficient. When $n = 9$, $n - 1 = 8$, which is a power of 2. But Gauss discovered that no construction exists for the regular 9-gon. The reason is that 9 is not prime.[24] What about the next case, in which we solve a series of four quadratic equations? Now the degree $n - 1$ of the corresponding single equation is $2 \times 2 \times 2 \times 2 = 16$. So $n = 17$, and this is prime.

By this point Gauss must have known he was on to something, but there is a further technical point, possibly fatal. Gauss had convinced himself that in order for construction for a regular pentagon with a prime number of sides to exist, that prime must be a power of 2, plus 1. So this condition is necessary for a construction to exist: if it fails, there is no such construction. However, the condition might not be sufficient: in fact there are plenty of equations of degree 16 that do not reduce to a series of four quadratics.

There was a reason to be optimistic, however: the Greek constructions. Which primes occurred there? Only three: 2, 3, and 5. These are all 1 more than a power of 2, namely $2^0 + 1$, $2^1 + 1$, and $2^2 + 1$. The algebra associated with the pentagon provides further clues. Thinking it all through, Gauss proved that the degree-16 polynomial associated with the 17-sided polygon can indeed be reduced to a series of quadratics. Therefore a ruler-and-compass construction must exist. A similar method proved that the same is true whenever the number of sides is a prime that is 1 greater than some power of 2. The ideas are a tribute to Gauss's ability to understand mathematical patterns. At their heart are some general theorems in number theory, which I won't go into here. The point is, none of this was accidental. There were solid structural reasons for it to work. You just had to be a Gauss to notice them.

Gauss didn't provide an explicit construction, but he did give a formula for the solutions of the degree-16 equation that can be turned into such a construction if you really want one.[25] When he wrote down his ideas in the *Disquisitiones Arithmeticae*, he omitted quite a few of the details, but he did assert that he possessed complete proofs. His epic discovery convinced him that he should devote his life to mathematics rather than languages. The Duke continued to support Gauss financially, but Gauss wanted something more permanent and reliable. When the astronomer Giuseppe Piazzi discovered the first asteroid, Ceres, only a few observations could be made before this new world became invisible against the glare of the Sun. Astronomers were worried that they wouldn't be able to find it again. In a *tour de force* that involved new techniques for calculating orbits, Gauss predicted where it would reappear – and he was right. This led to him being appointed Professor of Astronomy and Director of the Göttingen Observatory. He continued to hold the post for the rest of his life.

It turns out that 17 is not the only new number of this kind. Two more are known: $2^8 + 1 = 257$ and $2^{16} + 1 = 65,537$. (A bit of algebra shows that the power of 2 that occurs must itself be a power of 2; if not, then the number cannot be prime.) However, the pattern stops at that point, because $2^{32} + 1 = 4,294,967,297$ is equal to $641 \times 6,700,417$, hence is not prime. The so-called Fermat numbers $2^{2^n} + 1$ are known not to be prime for $n = 5, 6, 7, \ldots$ up to 32. Many larger Fermat numbers are also known not to be prime. No further prime Fermat numbers have been found, but their existence is by no means impossible.[26] A construction for the 257-sided polygon is known. One mathematician devoted many years to the 65,537-sided polygon, a somewhat pointless task, and his results contain errors anyway.[27]

The upshot of Gauss's analysis is that a regular polygon can be constructed with ruler and compass if and only if the number of sides is a product of a power of 2 and *distinct* odd prime Fermat numbers. In particular, a regular 9-sided polygon cannot be constructed in this manner. This immediately implies that at least one angle cannot be trisected, because the angle in an equilateral triangle is 60 degrees, and one third of that is 20 degrees. Given this angle, it is easy to construct a regular 9-sided polygon. Since that is impossible, there is no general ruler-and-compass construction to trisect an angle.

Gauss omitted many details of the proofs when he wrote up his results, and mathematicians couldn't simply take his word for it. In 1837 the French mathematician Pierre Wantzel published a complete proof of Gauss's characterisation of constructible regular polygons, and deduced the impossibility of trisecting a general angle by ruler-and-compass construction. He also proved that it is impossible to construct a cube whose volume is twice that of a given cube, another ancient Greek problem known as 'duplicating the cube'.

Both angle-trisection and cube-duplication turn out to be impossible because the lengths involved satisfy irreducible *cubic* equations – degree 3. Since 3 is not a power of 2, this knocks the question on the head. However, this method didn't seem to work for the problem of squaring the circle, for interesting reasons. A circle of unit radius has area π, and square of that area has side $\sqrt{\pi}$. Geometric constructions for square roots exist, and so do constructions for squares, so squaring the circle boils down to starting with a line of length 1 and constructing one of length π. If π happened to satisfy an irreducible cubic equation – or any irreducible equation whose degree is not a power of 2 – then Wantzel's methods would prove that it's impossible to square the circle.

However, no one knew any algebraic equation that was satisfied exactly by π, let alone one whose degree is not a power of 2. The school value 22/7 satisfies $7x - 22 = 0$, but that's just an approximation to π, a tiny bit too large, so it doesn't help. If it could be proved that no such equation exists – and many suspected this on the grounds that it would have been found if it did – the impossibility of squaring the circle would follow. Unfortunately, no one could prove that there is no such equation. The algebraic status of π was in limbo. The eventual solution employed methods that did not just go beyond geometry: they went beyond algebra as well.

To appreciate the main issue here, we need to start with a simpler idea. There is an important distinction in mathematics between numbers that can be expressed as exact fractions p/q, where p and q

are whole numbers, and those that can't be so expressed. The former are said to be rational (they are ratios of whole numbers), and the latter are irrational. The approximation 22/7 to π is rational, for example. There are better approximations; a famous one is 355/113, correct to six decimal places. However, it is known that no fraction can represent π exactly: it is irrational. This long-suspected property was first proved by the Swiss mathematician Johann Heinrich Lambert in 1768. His proof is based on a clever formula for the tangent function in trigonometry, which he expressed as a continued fraction: an infinite stack of ordinary fractions.[28] In 1873 Charles Hermite found a simpler proof, based on formulas in calculus, which went further: it proved that π^2 is irrational. Therefore π is not the square root of a rational number either.

Lambert suspected something much stronger. In the article that proved π to be irrational, he conjectured that π is transcendental; that is, π does not satisfy any polynomial equation with integer coefficients. It transcends algebraic expression. Subsequent discoveries proved him right. The breakthrough came in two stages. Hermite's new method for proving irrationality set the scene by hinting that calculus – more precisely, its rigorous version, analysis – might be a useful strategy. By pushing that idea further, Hermite found a wonderful proof that the other famous curious number in mathematics, the base e of natural logarithms, is transcendental. Numerically, e is roughly 2.71828, and if anything it is even more important than π. Hermite's transcendence proof is magical, a rabbit extracted with a flourish from the top hat of analysis. The rabbit is a complicated formula related to a hypothetical algebraic equation that e is assumed to satisfy. Using algebra, Hermite proves that this formula is equal to some nonzero integer. Using analysis, he proves that it must lie between $-\frac{1}{2}$ and $\frac{1}{2}$. Since the only integer in this range is zero, these results are contradictory. Therefore the assumption that e satisfies an algebraic equation must be false, so e is transcendental.

In 1882 Ferdinand Lindemann added some bells and whistles to Hermite's method, and proved that if a nonzero number satisfies an algebraic equation, then e raised to the power of that number does not satisfy an algebraic equation. He then took advantage of a relationship that was known to Euler involving π, e, and the imaginary number i: the famous formula $e^{i\pi} = -1$. Suppose that π satisfies some algebraic equation. Then so does iπ, and Lindemann's theorem implies that -1 does *not* satisfy an algebraic equation. However, it visibly does: it is the solution of $x + 1 = 0$. The only way out of this logical contradiction is that π does not satisfy an algebraic equation; that is, it is transcendental. And that means you can't square the circle.

It was a long and indirect journey from Euclid's geometry to Lindemann's proof, and it took more than 2000 years, but mathematicians finally got there. The story doesn't just tell us that the circle can't be squared. It is an object lesson in how great mathematical problems get solved. It required mathematicians to formulate carefully what they meant by 'geometric construction'. They had to pin down general features of such constructions that might place limits on what they could achieve. Finding those features required making connections with another area of mathematics: algebra. Solving the algebraic problem, even in simpler cases such as the construction of regular polygons, also involved number theory. Dealing with the difficult case of π required further innovations, and the problem had to be transported into yet another area of mathematics: analysis.

None of these steps was simple or obvious. It took about a century to complete the proof, even when the main ideas were in place. The mathematicians involved were among the best of their age, and at least one was among the best of *any* age. Solving great problems requires a deep understanding of mathematics, plus persistence and ingenuity. It can involve years of concentrated effort, most of it apparently fruitless. But imagine how it must feel when your persistence pays off, and you crack wide open something that has baffled the rest of humanity for centuries. As President John F. Kennedy said in 1962 when announcing the Moon landing project: 'We choose to ... do [these] ... things, not because they are easy, but because they are hard.'

Few stories in mathematics end, and π is no exception. Every so often, amazing new discoveries about π appear. In 1997 Fabrice Bellard announced that the trillionth digit of π, in binary notation, is 1.[29] What made the statement remarkable was not the answer. The amazing feature was that he didn't calculate any of the earlier digits. He just plucked one particular digit from the air.

The calculation was made possible by a curious formula for π discovered by David Bailey, Peter Borwein, and Simon Plouffe in 1996. It may seem a bit complicated, but let's take a look anyway:

$$\pi = \sum_{n=0}^{\infty} \frac{1}{2^{4n}} \left(\frac{4}{8n+1} - \frac{2}{8n+4} - \frac{1}{8n+5} - \frac{1}{8n+6} \right)$$

The big Σ means 'add up', over the range specified. Here n runs from 0 to infinity (∞). Bellard actually used a formula that he had derived using similar methods, which is marginally faster for computations:

$$\pi = \frac{1}{64} \sum_{n=0}^{\infty} \frac{(-1)^n}{2^{10n}} \left(-\frac{32}{4n+1} - \frac{1}{4n+3} + \frac{256}{10n+1} - \frac{64}{10n+3} \right.$$
$$\left. - \frac{4}{10n+5} - \frac{4}{10n+7} + \frac{1}{10n+9} \right)$$

The key point is that many of the numbers that occur here – 1, 4, 32, 64, 256, and also 2^{4n} and 2^{10n} – are powers of 2, which of course are very simple in the binary system used for the internal workings of computers. This discovery stimulated a flood of new formulas for π, and for several other interesting numbers. The record for finding a single binary digit of π is broken regularly: in 2010 Yahoo's Nicholas Sze computed the two-quadrillionth binary digit of π, which turns out to be 0.

The same formulas can be used to find isolated digits of π in arithmetic to the bases 4, 8, and 16. Nothing of the kind is known for any other base; in particular, we can't compute decimal digits in isolation. Do such formulas exist? Until the Bailey-Borwein-Plouffe formula was found, no one imagined it could be done in binary.

4

Mapmaking mysteries
Four Colour Theorem

MANY OF THE GREAT MATHEMATICAL problems stem from deep and difficult questions in well-established areas of the subject. They are the big challenges that emerge when a major area has been thoroughly explored. They tend to be quite technical, and everyone in the area knows they're hard to answer, because many experts have tried and failed. The area concerned will already possess many powerful techniques, massive mathematical machines whose handles can be cranked if you've done your homework – but if the problem is still open, then all of the plausible ways to use those techniques have already been tried, and *they didn't work*. So either there is a less plausible way to use the tried-and-tested techniques of the area, or you need new techniques.

Both have happened.

Other great problems are very different. They appear from nowhere – a scribble in the sand, a scrawl in a margin, a passing whim. Their statements are simple, but because they do not already have an extensive mathematical background, there are no established methods for thinking about them. It may take many years before their difficulty becomes apparent: for all anyone knows, there might be some clever but straightforward trick that solves them in half a page. The four colour problem is of this second kind. It took decades before mathematicians began to grasp how difficult the question was, and for a large part of that time they thought that it *had* been solved, in a few pages. It seemed to be a fringe issue, so few people bothered to take it seriously. When they did, the alleged solution turned out to be flawed. The final solution fixed the flaws, but by then the argument had become so complicated that massive computer assistance was required.

In the long run, both types of problem converge, despite their different backgrounds, because resolving them requires new ways of thinking. Problems of the first type may be embedded in a well-understood area, but the traditional methods of that area are inadequate. Problems of the second type don't belong to any established area – in fact, they motivate the creation of new ones – so there are no traditional methods that can be brought to bear. In both cases, solving the problem demands inventing new methods and forging new links with the existing body of mathematics.

We know exactly where the four colour problem came from, and it wasn't mathematics. In 1852 Francis Guthrie, a young South African mathematician and botanist working for a degree in law, was attempting to colour the counties in a map of England. He wanted to ensure that any two adjacent counties were assigned different colours, so that the borders were clear to the eye. Guthrie discovered that he needed only four different colours to complete the task, and after some experimentation convinced himself that this statement would be true for any map whatsoever. By 'adjacent' he meant that the counties concerned shared a border of nonzero length: if two counties touched at a point, or several isolated points, they could if necessary have the same colour. Without this proviso, there is no limit to the number of colours, because any number of regions can meet at a point, Figure 8 (left).

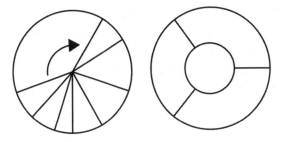

Fig 8 *Left*: Any number of regions can meet at a point. *Right*: At least four colours are necessary.

Wondering whether this statement was a known mathematical theorem, he asked his brother Frederick, who was studying mathematics under the distinguished but eccentric Augustus De Morgan at University College, London. De Morgan didn't know, so he wrote to an even more distinguished mathematican, the Irishman Sir William Rowan Hamilton:

> A student of mine [identified later as Frederick Guthrie] asked me today to give him a reason for a fact which I did not know was a fact – and do not yet. He says that if a figure be any how divided and the compartments differently coloured so that figures with any portion of common boundary *line* are differently coloured – four colours may be wanted but not more... Query cannot a necessity for five or more be invented... What do you say? And has it, if true, been noticed?

Frederick later referred to a 'proof' that his brother had suggested, but he also said that the key idea was a drawing equivalent to Figure 8, which proves only that fewer than four colours won't work.

Hamilton's reply was brief and unhelpful. 'I am unlikely to attempt your "quaternion" of colours very soon,' he wrote. At the time he was working on an algebraic system that became a lifelong obsession, analogous to the complex numbers but involving four types of number rather than the two (real and imaginary) of the complex numbers. This he called 'quaternions'. The system remains important in mathematics; indeed, it is probably more important now than it was in Hamilton's day. But it has never really attained the heights that Hamilton hoped for. Hamilton was just making an academic joke when he used the word, and for a long time there seemed to be no link between quaternions and the four colour problem. However, there is a reformulation of the problem that can be viewed as a statement about quaternions, so Hamilton's joke has a sting in its tail.[30]

Being unable to find a proof, De Morgan mentioned the problem to his mathematical acquaintances in the hope that one of them might come up with an idea. In the late 1860s the American logician, mathematician, and philosopher Charles Sanders Peirce claimed he had solved the four colour problem, along with similar questions about maps on more complex surfaces. His alleged proof was never published, and it is doubtful that the methods available to him would have been adequate.

Although the four colour problem is ostensibly about maps, it has no useful applications to cartography. Practical criteria for colouring maps mainly reflect political differences, and if that means that adjacent regions must have the same colour, so be it. The problem's interest lay entirely within pure mathematics, in a new area that had only just started to develop: topology. This is 'rubber-sheet geometry' in which shapes can be deformed in any continuous manner. But even there, the four colour problem didn't belong to the mainstream. It seemed to be no more than a minor curiosity.

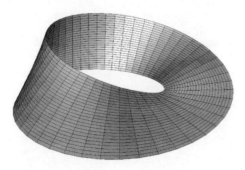

Fig 9 The Möbius band has only one side.

One of the pioneers of topology was August Möbius, famous today for his one-sided band, Figure 9. You can make a model by taking a strip of paper, bending it round into a ring like a short fat cylinder, twisting one end through 180 degrees, and gluing the ends together. A friend of his, the linguist Benjamin Weiske, set Möbius a puzzle: could an Indian king with five sons, all princes, divide up his kingdom so that the region belonging to each prince shared a border of nonzero length with the regions belonging to the other four princes? Möbius passed the puzzle on to his students as an exercise. But in the next lecture, he apologised for asking them to perform the impossible. By this he meant that he could *prove* it was impossible.[31]

It's hard to tackle this puzzle geometrically, because the shapes of the regions and how they are arranged might in principle be very complicated. Progress depends on a big simplification: all that really matters is which regions are adjacent to which, and how the common boundaries are arranged relative to each other. This is topological information, independent of the precise shapes. It can be represented in a clean, simple way, known as a graph – or, nowadays, a network, which is a more evocative term.

A network is a devastatingly simple concept: a set of vertexes, represented by dots, some of which are linked together by edges, drawn as lines. Take any map, such as Figure 10 (left). To convert it into a network, place a dot inside each region, Figure 10 (middle). Whenever two regions have a common length of border, draw a line between the corresponding dots, passing through that segment. If there are several separate common border segments, each gets its own line. Do this for all regions and all common border segments, in such a way that the lines do not cross each other, or themselves, and they meet only at dots. Then throw away the original map and retain only the dots and lines. These form the dual network of the map, Figure 10 (right).[32]

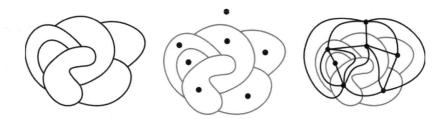

Fig 10 *Left*: A map. *Middle*: Place a point in each region. *Right*: Connect dots across borders to form the dual network (black lines and dots only).

The word 'dual' is used because the procedure takes regions, lines, and dots (junctions between regions of the map) and turns them into dots, lines, and regions. A region in the map corresponds to a dot in the dual network. A border segment in the map corresponds to a line in the dual network; not the same line, but one that crosses the border and links the corresponding dots. A point in the map where three or more regions meet corresponds to a region in the dual network bounded by a closed loop of lines. So the dual network is itself a map, because the lines enclose regions, and it turns out that the dual of the dual is the original map, give or take a few technicalities that exclude unnecessary dots and lines.

The problem of the five princes can be reinterpreted using the dual network: Is it possible to join five points in the plane by lines, with no crossings? The answer is 'no', and the key is Euler's formula, which states that if map on the plane consists of F faces (regions), E edges (lines), and V vertexes (dots), then $F + V - E = 2$. Here we count the rest of the plane, outside the network, as one big region. This formula was one of the first hints that topological considerations could be worth investigating, and it reappears in chapter 10.

The proof that the Indian princes puzzle is impossible starts by assuming that a solution exists, and deduces a contradiction. Any solution will have $V = 5$, the number of dots. Since each pair of dots is joined by a line, and there are 10 pairs, $E = 10$. Euler's theorem implies that $F = E - V + 2 = 7$. The regions of the dual network are surrounded by closed loops of lines, and only one line joins any pair of points; therefore these loops must contain at least 3 lines. Since there are 7 regions, that makes at least 21 lines … except that every line is being counted twice because it separates 2 regions. So there are at least $10\frac{1}{2}$ lines. The number of lines is an integer, so in fact there must be at least 11 lines. However, we already know there are 10 lines. This is a logical contradiction, and it proves that no such network exists. The king cannot divide his land in the prescribed way.

The encouraging aspect of this argument is that elegant topological methods enable us to prove something specific and interesting about maps. However, contrary to a common misconception, which De Morgan seems to have shared, the impossibility of solving the puzzle of the five Indian princes does *not* prove the four colour theorem. A proof may be wrong even if its conclusion is correct, or not known to be incorrect. If somewhere in an alleged proof I encounter a triangle with four sides, I can stop reading, because the proof is wrong. It doesn't matter what happens after that, or what the conclusion is. Our answer to the Indian princes puzzle shows that one particular way to disprove the four colour theorem doesn't work. However, that doesn't imply that no *other* way to disprove it could work. Potentially, there might be many obstacles to 4-colouring a map (from now on I will use this term rather than the clumsy 'colour the map with four colours'). The existence of five regions all adjacent to each other is merely one such obstacle. For all we know, there might be a very complicated map with 703 regions, such that however you 4-colour 702 of them, the final region always needs a fifth colour. That region would have to adjoin at least four others, but that's entirely feasible, and it doesn't require an Indian prince arrangement. If a map like that existed, it would prove that four colours are not enough. Any proof has to rule out that kind of obstacle. And that statement is valid even if I don't show you – can't show you – an explicit example of such an obstacle.

For a time the four colour problem seemed to have sunk without trace, but it resurfaced in 1878 when Arthur Cayley mentioned it at a meeting of the London Mathematical Society. Despite its name, this organisation represented the whole of British (or at least English) mathematics, and its founder was De Morgan. Cayley asked whether anyone had obtained a solution. His query was published soon after in the science journal *Nature*. A year later he wrote a more extended article for the *Proceedings of the Royal Geographical Society*.[33] It presumably seemed a logical place to put the paper, because the problem is ostensibly about maps. He may even have been asked to submit it. But it wasn't really a sensible choice, because no mapmaker would have any reason to want to know the answer, other than idle curiosity. Unfortunately, the choice of journal meant that few mathematicians would be aware of the article's existence. That was a pity, because Cayley explained why the problem might be tricky.

In chapter 1, I said that a proof is a bit like a battle. The military recognises a difference between tactics and strategy. Tactics are how you win local skirmishes; strategy sets the broad structure of the campaign. Tactics involve detailed troop movements; strategy involves broad plans, with room for many different tactical decisions at any stage. Cayley's article was short on tactics, but it contained the vaguest hint of a strategy which, in the fullness of time, cracked the four colour problem wide open. He observed that adding regions one at a time didn't work if you followed the obvious line of reasoning. But maybe it would work if you found a less obvious line of reasoning.

Suppose you take a map, remove one region – say by merging it with a neighbour, or shrinking it to a point. Suppose that the resulting map can be 4-coloured. Now put the original region back again. If you're lucky, its neighbours might use only three colours. Then all you have to do is colour it using the fourth. Cayley's point was that this procedure might not work, because the neighbours of the final region might use four distinct colours. But that doesn't mean you're stuck. There are two ways to get round this obstacle: you may have chosen the wrong region, or you may have chosen the wrong way to colour the smaller map.

Still running on unsubstantiated suppositions (this is a very effective way to get research ideas, although at some point you have to substantiate them), assume that something of this kind can always be fixed up. Then it tells you that a map can always be 4-coloured provided some smaller map can be 4-coloured. This may not seem like progress: how do we know the smaller map can be 4-coloured? The answer is that the same procedure applies to the smaller map, leading to an even smaller map ... and so on. Eventually you get to a map so small that it has only four regions, and then you know it can be 4-coloured. Now you reverse your steps, colouring slightly larger maps at each stage ... and eventually climb back to your original map.

This line of reasoning is called 'proof by mathematical induction'. It's a standard method with a more technical formulation, and the logic behind it can be made rigorous. Cayley's proposed proof strategy becomes more transparent if the method is reformulated using a logically equivalent concept: that of a minimal criminal. In this context, a criminal is any hypothetical map that can't be 4-coloured. Such a map is minimal if any map with a smaller number of countries *can* be 4-coloured. If a criminal exists, there must be a minimal one: just choose a criminal with the smallest possible number of regions. Therefore, if minimal criminals do not exist, then criminals do not exist. And if there are no criminals, the four colour theorem must be true.

The induction procedure boils down to this. Suppose we can prove that 4-colouring a minimal criminal is always possible, provided some related smaller map can be 4-coloured. Then the minimal criminal can't actually be a criminal. Because the map is minimal, *all* smaller maps can be 4-coloured, so by what we've supposed can be proved, the same is true of the original map. Therefore there are no minimal criminals, so there are no criminals. This idea shifts the focus of the problem from all maps to just the hypothetical minimal criminals, and specifying a reduction procedure − a systematic way to turn a 4-colouring of some related smaller map into a 4-colouring of the original map.

Why fuss about with minimal criminals, rather than plain criminals? It's a matter of technique. Even though we initially don't know whether criminals exist, one of the paradoxical but useful features of this strategy is that we can say quite a lot about what minimal ones would look like if they did exist.

This requires the ability to think logically about hypotheticals, a vital skill for any mathematician. To give a flavour of the process, I'll prove the *six* colour theorem. To do so, we borrow a trick from the five princes puzzle, and reformulate everything in terms of the dual network, in which regions become dots. The four colour problem is then equivalent to a different question: Given a network in the plane whose lines do not cross, is it possible to 4-colour the *dots*, so that two dots joined by a line always have different colours? The same reformulation applies with any number of colours.

To illustrate the power of minimal criminals, I'm going to use them to prove that any planar network can be 6-coloured. Again the main technical tool is Euler's formula. Given a dot in the dual network, define its neighbours to be those dots that are linked to it by a line. A dot may have many neighbours, or just a few. It can be shown that Euler's formula implies that some dots must have few neighbours. More precisely, in a planar network it is impossible for all dots to have six or more neighbours. I've put a proof of this in the notes to avoid interrupting the flow of ideas.[34] This fact provides the necessary lever to start pulling the problem to pieces. Consider a hypothetical minimal criminal for the 6-colour theorem. This is a network that can't be 6-coloured, but every smaller network can be 6-coloured. I now prove this map can't exist. By the above consequence of Euler's formula, it contains at least one dot with five or fewer neighbours. Temporarily delete this dot and the lines that link it to its neighbours. The resulting network has fewer dots, so by minimality it can be 6-coloured. (Here is where we get stuck unless our hypothetical criminal is minimal.) Now put the deleted dot and lines back. That dot has at most five neighbours, so there is always a sixth colour. Use this to colour the deleted dot. Now we have successfully 6-coloured our minimal criminal – but that contradicts its criminality. So minimal criminals for the 6-colour theorem don't exist, and that implies that the six colour theorem is true.

This is encouraging. Until now, for all we knew, some maps might need 20 colours, or 703, or millions. Now we know that maps like that are no more real than the pot of gold at the end of the rainbow. A specific, limited number of colours definitely works for *any* map. This is a genuine triumph for minimal criminals, and it encouraged mathematicians to tighten up the argument in the hope of replacing six colours by five, or if you were really clever, four.

All criminals need lawyers. A barrister named Alfred Kempe was at the meeting in which Cayley mentioned the four colour problem. He had studied mathematics under Cayley as an undergraduate at Cambridge, and his interest in the subject remained undiminished. Within a year, Kempe convinced himself that he had cracked the problem, and he published his solution in 1879 in the newly founded *American Journal of Mathematics*. A year later he published a simplified proof, which corrected some errors in the first one. He pointed out that

> A very small alteration in one part of a map may render it necessary to recolour it throughout. After a somewhat arduous search, I have succeeded ... in hitting upon the weak point, which proved an easy one to attack.

I'll reinterpret Kempe's ideas in terms of the dual network. Once again he started from Euler's formula and the consequent existence of a dot with either three, four, or five neighbours. (A dot with two neighbours lies in the middle of a line, and contributes nothing to the network or the map: it can safely be omitted.)

If there is a dot with three neighbours, the procedure I used in proving the six colour theorem applies when there are only four colours. Remove the dot and the lines that meet it, 4-colour the result, put the dot and lines back, use a spare colour for the dot. We may therefore assume that no dot has three neighbours.

If there is a dot with four neighbours the above tactic fails, because a spare colour might not be available. Kempe devised a clever way to deal with this obstacle: delete that dot anyway, but after doing so, change the colouring of the resulting smaller map so that two of those four neighbours have the same colour. After this change, the neighbours of the deleted dot use at most three colours, leaving a spare one for the deleted dot. The basic idea of Kempe's recolouring scheme is that two of the neighbouring dots must have different colours – say red and blue, with the other colours being green or yellow. If both are green or both yellow, the other colour is available for the deleted dot. So we can assume one is green and one yellow. Now find all of the dots that can be connected to the blue one by a sequence of lines, using only blue and red dots. Call this a blue-red Kempe chain.[35] By definition, every neighbour of any dot in the Kempe chain, that is not itself in the Kempe chain, is either green or yellow, because a blue or red neighbour would already be in the chain. Having found such a chain, observe that swapping the two colours blue and red for all dots within the chain produces another colouring of the network, still satisfying the key condition that adjacent dots get different colours, Figure 11.

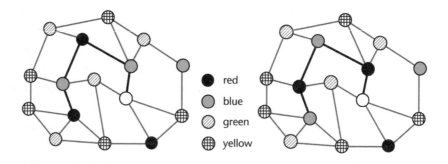

Fig 11 Swapping colours in a Kempe chain (thick black lines) associated with a dot of degree 4 (white) that has neighbours of all four colours. *Left*: Original colours. *Right*: With swapped colours, blue is available for the white dot.

If the red neighbour of our original dot is not in this blue-red chain, make such a change. The blue neighbour of the original dots turns red; the red neighbour remains red. Now the neighbours of the original dot use at most three different colours: red, green, and yellow. That leaves blue for the original dot, and we're done. However, the blue-red chain might loop round and join up with the blue neighbour. If so, leave the blue and red chain alone, and use the same trick for the yellow and green neighbours of the original dot instead. Start with the green one and form a green-yellow Kempe chain. This chain *can't* join up with the yellow neighbour, because the previous blue-red chain gets in the way. Swap yellow and green, and you're done.

That leaves one final case, when there are no dots with three or four neighbours, but at least one has five neighbours. Kempe proposed a similar but more complicated recolouring rule, which seemed to sort out that case as well. Conclusion: the four colour theorem is true, and Kempe had proved it. It even hit the media: *The Nation*, an American magazine, mentioned the solution in its review section.

Kempe's proof seemed to have laid the problem to rest. For most mathematicians it was a done deal. Peter Guthrie Tait continued publishing papers on the problem, seeking a simpler proof; this led him to some useful discoveries, but the simpler proof eluded him.

Enter Percy Heawood, a lecturer in mathematics at Durham University known by the nickname 'Pussy', thanks to his magnificent moustache. As an undergraduate at Oxford he had learned of the four colour problem from Henry Smith, the Professor of Geometry. Smith told him that the theorem, though probably true, was unproved, so Heawood had a go. Along the way he came across Kempe's paper, and tried to understand it. He published the outcome in 1889 as 'Map colour theorem', regretting that the aim of his article was more 'destructive than constructive, for it will be shown that there is a defect in the now apparently recognized proof.' Kempe had made a mistake.

It was a subtle mistake, and it occurred in the recolouring method when the dot being deleted had five neighbours. Kempe's scheme could occasionally change the colour of some dot as a knock-on effect of later changes. But Kempe had assumed that once the colour of a dot had been changed, it didn't change again. Heawood found a network for which Kempe's recolouring scheme went wrong, so his proof was flawed. Kempe was quick to acknowledge the error, and added that he had 'not succeeded in remedying the defect'. The four colour theorem was up for grabs again.

Heawood extracted some crumbs of comfort for Kempe from the débâcle: his method successfully proved the five colour theorem. Heawood also worked on two generalisations of the problem: empires, in which the regions can consist of several disconnected pieces, all requiring the same colour; and maps on more complicated surfaces. The analogous question on a sphere has the same answer as it does for the plane. Imagine a map on a sphere, and rotate it until the north pole is somewhere inside one region. If you delete the north pole you can open up the punctured sphere to obtain a space that is topologically equivalent to the infinite plane. The region that contains the pole becomes the infinitely large one surrounding the rest of the map. But there are other, more interesting, surfaces. Among them is the torus, shaped like a doughnut with a hole, Figure 12 (left).

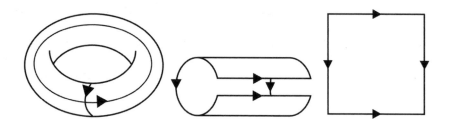

Fig 12 Cutting a torus open and unwrapping it to make a square.

There is a useful way to visualise the torus, which often makes life simpler. If we cut the torus along two closed curves, Figure 12 (middle), we can open it out into a square, Figure 12 (right). This transformation changes the topology of the torus, but we can get round that by 'identifying' opposite edges of the square. In effect (and a rigorous definition makes this idea precise) we agree to treat corresponding points on these edges as though they were identical. To see how this goes, reverse the sequence of pictures. The square gets rolled up, and opposite edges really do get glued together. Now comes the clever part. You don't actually need to roll up the square and join corresponding edges. You can just work with the flat square, provided you bear in mind the rule for identifying the edges. Everything you do to the torus, such as drawing curves on it, has a precise corresponding construction on the square.

Heawood proved that seven colours are both necessary and sufficient to colour any map on a torus. Figure 13 (left) shows that seven are necessary, using a square to represent the torus as just described. Observe how the regions match at opposite edges. There are surfaces like a torus, but with more holes, Figure 13 (right). The number of holes is called the genus, and denoted by the letter g. Heawood conjectured a formula for the number of colours required on

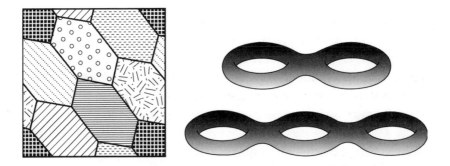

Fig 13 *Left*: Seven-colour map on a torus. The torus is represented as a square whose opposite sides are conceptually 'wrapped round' so that they glue together. Regions of the map are required to match up across corresponding edges. *Right*: Toruses with two and three holes.

a torus with g holes when $g \geqslant 1$: it is the smallest whole number less than or equal to

$$\frac{7 + \sqrt{48g + 1}}{2}$$

When g ranges from 1 to 10, this formula yields the numbers

7 8 9 10 11 12 12 13 13 14

The number of colours specified by the formula grows more slowly than the genus, and often it makes no difference if you add an extra hole to the torus. This is a surprise, because every extra hole provides more freedom to invent complicated maps.

Heawood didn't just pluck this formula from thin air. It arose by generalising the way I proved the six-colour theorem in the plane. He could prove that this number of colours is always sufficient. The big question, for many years, was whether this number can be made smaller. Examples for small values of the genus suggested that Heawood's estimate is the best possible. In 1968, after a lengthy investigation, Gerhard Ringel and John W.T. (Ted) Youngs filled in the final details in a proof that this is correct, building on their own work and that of several others. Their methods are combinatorial, based on special kinds of networks, and complicated enough to fill an entire book.[36]

When $g = 0$, that is, for maps on a sphere, Heawood's formula gives four colours, but his proof of sufficiency doesn't work on a sphere. Despite impressive progress for surfaces with at least one hole, the original four colour problem was still up for grabs. The few mathematicians who were willing to devote serious efforts to the question settled in for what, in warrior terms, was likely to be a long siege. The problem was a heavily defended castle; they hoped to build ever more powerful siege engines, and keep knocking pieces off until the castle walls fell. And they did, and it didn't. However, the attackers slowly accumulated a lot of information about how not to solve the problem, and the types of obstacle that seemed unavoidable. Out of these failures an ambitious strategy began to emerge. It was a natural extension of Kempe's and Heawood's methods, and it came in three parts. I'll state them using the dual network, the standard viewpoint nowadays:

1   Consider a minimal criminal.

2   Find a list of unavoidable configurations: smaller networks, with the property that any minimal criminal must contain something in the list.

3   Prove that each of the unavoidable configurations is reducible. That is: if a smaller network, obtained by deleting the unavoidable configuration, can be 4-coloured, then these colours can be redistributed so that when the unavoidable configuration is restored, the 4-colouring of the smaller network extends to the whole network.

Putting these three steps together, we can prove that a minimal criminal doesn't exist. If it did, it would contain an unavoidable configuration. But the rest of the network is smaller, so minimality implies that it can be 4-coloured. Reducibility now implies that the original network can be 4-coloured. This is a contradiction.

In these terms, Kempe had correctly found a list of unavoidable configurations: a dot with three lines sticking out, one with four, and one with five, Figure 14. He had also correctly proved that the first two are reducible. His mistake lay in his proof that the third configuration is reducible. It's not. Proposal: replace this bad configuration by a longer list, making sure that the list remains unavoidable. Do this in

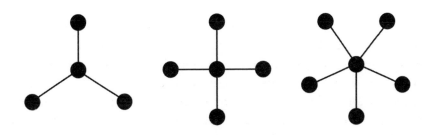

Fig 14 Kempe's list of unavoidable configurations.

such a manner that each configuration on the new list is reducible. That is: look for an unavoidable list of reducible configurations. If you succeed, you've proved the four colour theorem.

There might not be such a list, but this strategy is worth a try, and no one had any better ideas. It has one awkward inner tension, though. On the one hand, the longer the list, the more chance it has of being unavoidable, which is good. On the other hand, the longer the list, the less likely it is that every configuration in it will be reducible. If even a single one of them is not, the whole proof collapses, and this danger becomes more acute as the list grows. Which is bad. On the *third* hand ... a longer list provides more opportunities to choose reducible configurations, which is good. On the fourth hand, it increases the work needed to prove reducibility, which is bad. And on the fifth hand, there were no good methods for doing that anyway, which was worse.

This sort of thing is what makes great problems great.

So for a time the odd bit of the castle occasionally got chipped off, and its loss made not one whit of difference to the stronghold's solidity; meanwhile mainstream mathematics yawned, if it noticed at all. But someone was building a better battering-ram, and his name was Heinrich Heesch. His big contribution was a systematic way to prove that a configuration is reducible. He called it 'discharging', and it is roughly analogous to imagining that the dots in the network carry electric charges, and allowing the electricity to flow from one dot to another.

Even with this method, finding an unavoidable set of reducible configurations by hand would be a daunting task. The individual configurations would probably be fairly small, but there would have to be a lot of them. Heesch persevered, and in 1948 he gave a course of lectures suggesting that about 10,000 configurations would be needed. By then he had already proved that 500 candidates were reducible. In the audience was a young man named Wolfgang Haken, who later said that he hadn't actually understood much of Heesch's lectures at the time, but some of the main points had stuck in his head. Haken went on to study topology, making a major breakthrough in knot theory. This encouraged him to work on the Poincaré conjecture, chapter 10. For a particular line of attack, he classified the possibilities into 200 cases, solved 198 of them, and grappled with the remaining two for 13 years. At that point he gave up, and started working on the four colour problem instead. Haken clearly liked tough problems, but he was worried that something similar might happen with Heesch's 10,000 configurations. Imagine dealing successfully with 9,998, and getting stuck on the final two. So in 1967 he invited Heesch to visit the University of Illinois, where Haken was based, to ask his advice.

In those days computers were starting to become useful for real mathematics, but they were huge machines located in some central building, not things that sat on your desk or inside your briefcase. Haken wondered whether they could help. Heesch had already had the same idea and made a rough estimate of the complexity of the problem. This indicated that the best computer available to him wasn't up to the task. Illinois had a far more powerful supercomputer, ILLIAC-IV, so Haken requested time on it. But the supercomputer wasn't ready, so he was told to try the Cray 6600 in the Brookhaven Laboratory on Long Island. The director of the lab's computer centre was Yoshio Shimamoto, who had long been fascinated by the four colour problem – a stroke of good luck that gave Heesch and Haken access to the machine.

The computer lived up to expectations, but Haken started to wonder whether it could be used more efficiently. They were generating lots of reducible configurations and hoping to assemble an unavoidable list, but that strategy wasted a lot of time on potential configurations that turned out not to be reducible. Why not do it the other way round: make unavoidability the main objective, and check reducibility later? Of course, you'd need to use configurations that had a good chance of being reducible, but it seemed a better way to go. By then, however, the

Cray at Brookhaven was being used for more important things. Worse, several experts told Haken that the methods he wanted to use couldn't be turned into computer programs at all. He believed them, and gave a lecture saying that the problem couldn't be solved without computers, but now it seemed it couldn't be solved with computers either. He had decided to give up.

In the audience was an expert programmer, Kenneth Appel, who told Haken that the alleged experts were probably just trying to put him off because the programs would take a lot of work and the outcome was very uncertain. In Appel's view, there was no mathematical problem that could not be programmed. The crucial issue was whether the program would get anywhere in a reasonable time. They joined forces. The strategy evolved as improvements to the discharging method caused changes to the program and improvements to the program caused changes to the discharging method. This led them to a new concept: 'geographically good' configurations, which didn't contain certain nasty configurations that prevented reducibility. The chance of such a configuration being reducible was much improved, and the defining property was easy to check. Appel and Haken decided to prove theoretically, rather than by computer, that there was an unavoidable list of geographically good configurations. By 1974 they had succeeded.

This was encouraging, but they knew what was likely to happen. Some of their geographically good configurations would turn out not to be reducible, so they would have to remove those and replace them by a longer, more complicated, list. The calculation would be chasing its tail, and would succeed only if it caught it. Rather than wasting years on a fruitless pursuit, they did some rough calculations to estimate how long the process might take. The results were mildly encouraging, so the work went on. Theory and computation fed off and changed each other. At times, the computer seemed to have a mind of its own, 'discovering' useful features of configurations. Then the university administration bought itself a new, very powerful computer – more powerful than those available to the university's scientists. After some protest, and pointed questions, half of the machine's time was made available for scientific use. Appel and Haken's ever-changing list of unavoidable configurations stabilised at around 2000 of them. In June 1976, the computer ground out its last reducibility check, and the proof was complete. The story hit the media, starting in *The Times* and rapidly spreading worldwide.

They still had to make sure there were no silly mistakes, and by then several other teams were in hot pursuit. By July, Appel and Haken were confident their method worked, and they announced their proof officially to the mathematical community by circulating a preprint – a cheaply duplicated draft version of a paper, intended for later publication. At the time, it typically took between one and two years to get a mathematical paper into print. To avoid holding up progress the profession had to find a quicker way to get important results out to the community, and preprints were how they did it. Nowadays, the preprint goes on the web. Preprints are always provisional; full publication requires peer review. Preprints assist this process, because anyone can read them, look for mistakes or improvements, and tell the authors. In fact, the published version often differs considerably from the preprint, for just that reason.

The final proof took a thousand hours of computer time and involved 487 discharging rules; the results were published as two papers with a 450-page supplement showing all 1482 configurations. At the time, it was a *tour de force*.

The main reaction from the wider mathematical community, however, was vague disappointment. Not at the result; not at the remarkable computational achievement. What was disappointing was the method. In the 1970s, mathematical proofs were things that you put together by hand and checked by hand. As I said in Chapter 1, a proof is a story whose plot convinces you the statement is true. But this story didn't have a plot. Or if it did, there was a big hole in the middle:

> Once upon a time there was a beautiful conjecture. Her mother told her never to enter the dark, dangerous forest. But one day Little Four Colour Conjecture slipped away and wandered into the unavoidable forest. She knew that if each configuration in the forest were reducible, she would have a proof, become Little Four Colour Theorem, and be published in a journal run by Prince Ton. She came across a candy-coated computer, deep in the forest, and inside was a Wolf disguised as a programmer. And the Wolf said 'Yes, they are all reducible', and they all lived happily ever after.

No, it doesn't work. I'm being flippant, but the hole in this fairytale is the same as the hole in the Appel-Haken proof, or at least, what most mathematicians considered to be the hole in the proof. *How do we know the Wolf is right?*

We run our own computer program and find out whether it agrees. But however many times we do that, it won't have the same ring of authenticity as, say, my proof that you can't cover a hacked chessboard with dominoes. You can't grasp it as a whole. You couldn't check all the calculations by hand if you lived for a billion years. Worse, you wouldn't believe the answer if you could. Humans make mistakes. In a billion years, they make a lot of mistakes.

Computers, by and large, don't. If a computer and a human both do a really complicated piece of arithmetic and disagree, the smart money is on the computer. But that's not certain. A computer that is functioning exactly as designed can make an error; for example, a cosmic ray can zip through its memory and change a 0 to a 1. You can guard against that by doing the computation again, but more seriously, designers can make mistakes. The Intel P5 Pentium chip had an error in its routines for floating-point arithmetic: if asked to divide 4195835 by 3145727, it responded with 1.33373, when the correct answer is 1.33382. Apparently, four entries in a table had been left out.[37] Other things that can go wrong include the computer's operating system and bugs in the user's program.

A lot of philosophical hot air has been expended on the proposition that the Appel-Haken computer-assisted proof changed the nature of 'proof'. I can see what the philosophers are getting at, but the concept of proof that working mathematicians use is not the one we teach to undergraduates in mathematical logic classes. And even when that more formal concept applies, nothing requires the logic of each step to be checked by a human. For centuries, mathematicians have used machines for routine arithmetic. And even if a human does go through a proof line by line, finding no mistakes, how do we know they haven't missed one? Perfect, unassailable logic is an ideal at which we aim. Imperfect humans do the best they can, but they can never remove every element of uncertainty.

In *Four Colours Suffice*, Robin Wilson put his finger on a key sociological aspect of the community's reaction:

The audience split into two groups: the over-forties could not be convinced that a proof by computer was correct, while the under-forties could not be convinced that a proof containing 700 pages of hand calculations could be correct.

If our machines are better at some things than we are, it makes sense to use machines. Proof *techniques* may change, but they do that all the time anyway: it's called 'research'. The concept of proof does not radically alter if some steps are done by a computer. A proof is a story; a computer-assisted proof is a story that's too long to be told in full, so you have to settle for the executive summary and a huge automated appendix.

Since Appel and Haken's pioneering work, mathematicians have become accustomed to computer assistance. They still *prefer* proofs that rely solely on human brainpower, but most of them don't make that a requirement any more. In the 1990s, though, there was still a certain amount of justifiable unease about the Appel-Haken proof. So instead of rechecking the work, some mathematicians decided to redo the whole proof, taking advantage of new theoretical advances and much improved computers. In 1994 Neil Robertson, Daniel Sanders, Paul Seymour, and Robin Thomas threw away everything in the Appel-Haken paper except the basic strategy. Within a year they had found an unavoidable set of 633 configurations, each of which could be proved reducible using just 32 discharging rules. This was much simpler than Appel and Haken's 1482 configurations and 487 discharging rules. Today's computers are so fast that the entire proof can now be verified on a home computer in a few hours.

That's all very well, but the computer is still king. Can we get rid of it? There is a growing feeling that in this particular instance a story that humans can grasp in its entirety may not be totally inconceivable. Perhaps new insights into the four colour problem will eventually lead to a simpler proof, with little or no computer assistance, so that mathematicians can read it, think about it, and say 'Yes!' We don't know such a proof yet, and it may not exist, but there's a feeling in the air...

Mathematicians are learning a lot about networks. Topologists and geometers are finding deep relations between networks and utterly different areas of mathematics, including some that apply to mathematical physics. One of the concepts that turns up, from time to time, is curvature. The name is apt: the curvature of a space tells you how bent it is. If it's flat like the plane, its curvature is zero. If it bends in the same direction, the way a hilltop curves downwards on all sides, it has positive curvature. If it's like a mountain pass, curving up in some directions but down in others, it has negative curvature. There are geometric theorems, descendants of Euler's formula, that relate networks drawn in a space to the space's own curvature. Heawood's formula for a g-holed torus hints at this. A sphere has positive curvature, a torus represented as a square with opposite edges identified, Figure 12 (right), has zero curvature, and a torus with two or more holes has negative curvature. So there is some sort of link between curvature and map-colouring.

Behind this link is a useful feature of curvature: it's hard to get rid of it. It's like a cat under a carpet. If the carpet is flat, there's no cat, but if you see a bump, there's a cat underneath. You can chase the cat around the carpet, but all that does is move the bump from one place to another. Similarly, curvature can be moved, but not removed. Unless the cat gets to the edge of the carpet, in which case it can escape, taking its curvature with it. Heesch's discharging rules look a bit like curvature in disguise. They shift electrical charge around, but don't destroy it. Might there exist some concept of curvature for a network, and some cunning discharging rule that, in effect, pushes curvature around?

If so, you might be able to persuade a network to colour itself automatically. Assign curvature to its dots (and perhaps lines); then let the network redistribute the curvature more evenly. Perhaps 'evenly' here implies that four colours suffice, if we set everything up correctly. It's only an idea, it's not mine, and I haven't explained it in enough detail to make much sense. But it reflects some mathematicians' intuition, and it offers hope that a more conceptual proof of the four colour theorem – a ripping yarn rather than a summary with a billion telephone books as an appendix – might yet be found. We will encounter a similar idea, in a far more sophisticated context, in chapter 10, and it solved an even greater problem in topology.

5

Sphereful symmetry
Kepler Conjecture

T ALL STARTED WITH A SNOWFLAKE.

Snow has a strange beauty. It falls from the sky in fluffy white flakes, it blows in the wind to create soft humps and hillocks that cover the landscape, it forms spontaneously into outlandish shapes. It's cold. You can ski on it, ride a sleigh over it, make snowballs and snowmen out of it ... and if you're unlucky, you can be buried by thousands of tonnes of it. When it goes away, it doesn't go back into the sky – not directly as white flakes. It turns into plain ordinary water. Which can evaporate and go back in to the sky, of course, but may travel down rivers to the sea along the way, and spend a very long time in the oceans. Snow is a form of ice, and ice is frozen water.

This is not news. It must have been obvious to Neanderthals.

Snowflakes are by no means formless lumps. When they are pristine, before they start to melt, many of them are tiny, intricate stars: flat, six-sided, and symmetric. Others are simple hexagons. Some have less symmetry, some have a substantial third dimension, but sixfold snowflakes are iconic and widespread. Snowflakes are ice crystals. That's not news either: you just have to recognise a crystal when you see one. But these are no ordinary crystals, with flat, polygonal facets. Their most puzzling feature adds a dash of chaos: despite having the same symmetry, the detailed structure differs from each snowflake to the next. No two snowflakes are alike, they say. I've always wondered how they know, but the numbers favour that view if you are sufficiently pedantic about what counts as being alike.

Why are snowflakes six-sided? Four hundred years ago, one of the great mathematicians and astronomers of the seventeenth century asked himself that question, and thought his way towards an answer. It turned out to be an amazingly good answer, all the more so because he carried out no special experiments. He just put together some simple ideas that were known to everyone. Like the way pomegranate seeds are packed inside the fruit.

His name was Johannes Kepler, and he had a very good reason to think about snowflakes. His livelihood depended on a rich sponsor, John Wacker of Wackenfels. At that time Kepler was court mathematician to the Holy Roman Emperor Rudolf II, and Wacker, a diplomat, was a counsellor to the emperor. Kepler wanted to give his sponsor a New Year present. Ideally it should be cheap, unusual, and stimulating. It should give Wacker an insight into the remarkable discoveries that his money was making possible. So Kepler collected his thoughts about snowflakes in a small book, and that was the present. Its title was *De Nive Sexangula* ('On the six-cornered snowflake'). The date was 1611. Tucked away inside it, one of the main steps in Kepler's thinking, was a brief remark: a mathematical puzzle that would not be solved for 387 years.

Kepler was an inveterate pattern-seeker. His most influential scientific work was the discovery of three basic laws of planetary motion, the first and best known being that the orbit is an ellipse. He was also a mystic, thoroughly immersed in the Pythagorean world-view that the universe is based on numbers, patterns, and mathematical shapes. He did astrology as well as astronomy: mathematicians often moonlighted as astrologers in those days, because they could actually do the sums to work out when Aquarius was in the ascendant. Wealthy patrons, even royalty, paid them to cast horoscopes.

In his book, Kepler pointed out that snow begins as water vapour, which is formless, yet somehow the vapour turns into six-sided solid flakes. Some agent must cause this transition, Kepler insisted:

> Did [this agent] stamp the six-cornered shape on the stuff as the stuff demanded, or out of its own nature – a nature, for instance, in which

there is inborn either the idea of the beauty inherent in the hexagon or knowledge of the purpose which that form subserves?

In search of the answer, he considered other examples of hexagonal forms in nature. Honeycombs in beehives sprang to mind. They are made from two layers of hexagonal cells, back to back, and their common ends are formed by three rhombuses – parallelograms with all sides equal. This shape reminded Kepler of a solid called a rhombic dodecahedron, Figure 15. It's not one of the five regular solids that the Pythagoreans knew about and Euclid classified, but it has one distinctive property: identical copies can pack together exactly to fill space, with no gaps. The same shape occurs in pomegranates, where small round seeds grow, get squeezed together, and are therefore forced to create an efficient packing.

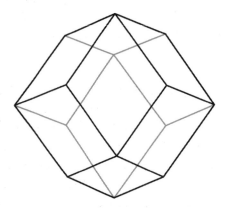

Fig 15 The rhombic dodecahedron, a solid with 12 rhombic faces.

Like any sensible mathematician, Kepler begins with the simplest case in which the spheres form a single plane layer. This is equivalent to packing identical circles in the plane. Here he finds just two regular arrangements. In one, the spheres are arranged in squares, Figure 16 (left); in the other, they are arranged in equilateral triangles, Figure 16 (right). These arrangements, repeated over the entire infinite plane, are the square lattice and the triangular lattice. The word 'lattice' refers to their spatially periodic pattern, which repeats in two independent directions. The figures necessarily show a finite portion of the pattern, so you should ignore the edges. The same goes for Figures 17–20 below. Figure 16 (left) and (right) both show five rows of spheres, and in each row they touch their neighbours. However, the triangular lattice is slightly squashed: its rows are closer together. So the spheres in the triangular lattice are more closely packed than those in the square lattice.

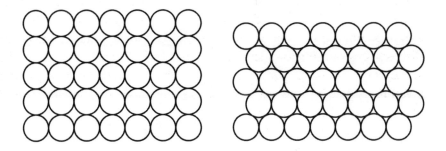

Fig 16 *Left*: Square lattice packing. *Right*: Triangular (also called hexagonal) lattice packing.

Next, Kepler asks how successive layers of this kind can be placed on top of each other, and he considers four cases. For the first two, all layers are square lattices. One way to stack the layers is to place the spheres in each layer directly above those below. Then every sphere will have six immediate neighbours: four within its layer, one above, and one below. This packing is like a three-dimensional chessboard made of cubes, and it would turn into that if you inflated the spheres until they could expand no further. But this, says Kepler, 'will not be the tightest pack'. It can be made tighter by sliding the second layer sideways, so that its spheres fit neatly into the dents between the spheres in the layer below, Figure 17 (left). Repeat this process, layer by layer, Figure 17 (right). Now each sphere has twelve neighbours: four in its own layer, four above, and four below. If you inflate them you fill the space with rhombic dodecahedrons.

In the other two cases, the layers are triangular lattices. If they are stacked so that the spheres in each layer lie directly above those below, then each sphere has eight neighbours: six in its own layer, one above, and one below. Alternatively, the spheres in the next layer can again be fitted into the dents between the spheres in the layer below. Now each

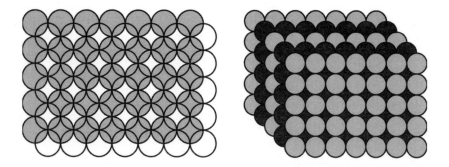

Fig 17 *Left*: Adding a second layer of spheres (open circles) on top of the first layer (grey). *Right*: Repeating this construction.

sphere has twelve neighbours: six in its own layer, three above, and three below. This is the same number of neighbours as the spheres in the second arrangement of square layers, and Kepler provides a careful analysis of the geometry to show that this fourth arrangement is actually the same as the second one. The only difference is that the square layers are no longer horizontal, but slanted at an angle. He writes: 'Thus in the closest pack in three dimensions, the triangular pattern, cannot exist without the square, and vice versa.' I'll come back to that: it's important.

Having sorted out the basic geometry of sphere packing, Kepler returns to the snowflake and its sixfold symmetry. He is reminded of the triangular lattice packing of spheres in a plane, in which each sphere is surrounded by six others, which form a perfect hexagon. This, he decides, must be why snowflakes are six-sided.

This chapter isn't primarily about snowflakes, but Kepler's explanation of their symmetry is very similar to the one we would offer today, so it would be a shame to stop here. Why are they – how *can* they be – so diverse, yet symmetric? When water crystallises to create ice, the atoms of hydrogen and oxygen that form the molecules of water pack together into a symmetric structure, the crystal lattice. This lattice is more complicated than any of Kepler's arrangements of spheres, but its dominant symmetry is sixfold. A snowflake grows from a tiny 'seed' with just a few atoms, arranged like a small piece of the lattice. This seed has the same sixfold symmetry, and it sets the scene for the growth of the ice crystal as the winds blow it this way and that inside a storm cloud.

The great variety of snowflake patterns is a consequence of changing conditions in the cloud. Depending on temperature and humidity, crystal growth may be uniform, with atoms being added at the same rate all along the boundary, leading to straight-sided hexagons, or it can be dendritic, with a growth rate that varies from place to place, resulting in treelike structures. As the growing flake is transported up and down through the cloud, these conditions keep changing, randomly. But the flake is so tiny that at any given moment, the conditions are essentially the same at all six corners. So they all do the same thing. Every snowflake carries traces of its history. In practice the sixfold symmetry is never exact, but it's often very close. Ice is strange stuff, and other shapes are possible too – spikes, flat plates, hexagonal prisms, prisms with plates on their ends. The full story is very complicated, but everything hinges on how the atoms in ice crystals are arranged.[38] In Kepler's day, atomic theory was at best a vague suggestion by a few ancient Greeks; it's amazing how far he got on the basis of folklore observations, thought experiments, and a feeling for pattern.

The Kepler conjecture is not about snowflakes as such. It is his offhand remark that stacking layers of close-packed spheres, so that successive layers fit into the gaps between those in the previous layer, leads to 'the closest pack in three dimensions'. The conjecture can be summarised informally: if you want to pack a large number of oranges into a big box, filling as much of the box as possible, then you should pack them together the way any greengrocer would.

The difficulty is not to find the *answer*. Kepler told us that. What's difficult is to prove he was right. Over the centuries, plenty of indirect evidence accumulated. No one could come up with a closer packing. The same arrangement of atoms is common in crystals, where efficient packing presumably corresponds to minimising energy, a standard principle that governs many natural forms. This kind of evidence was good enough to satisfy most physicists. On the other hand, no one could come up with a proof that there *wasn't* anything better. Simpler questions of the same kind, such as packing circles in the plane, turned out to have hidden depths. The whole area was difficult and full of surprises. All this worried mathematicians, even though most of them thought Kepler had the right answer. In 1958 C. Ambrose Rogers described the Kepler conjecture as something that 'many mathematicians believe, and all physicists know'.[39] This chapter describes how mathematicians turned belief into certainty.

To understand what they did, we need to take a close look at Kepler's arrangement of spheres, which is known as the face-centred cubic lattice. When we do, the subtleties of the problem start to emerge. The first question that springs to mind is why we use square layers. After all, the tightest pack in a single layer occurs for the *triangular* lattice. The answer is that we can also obtain the face-centred cubic lattice using triangular layers; this is the essence of Kepler's remark that 'the triangular pattern cannot exist without the square'. However, it is easier to describe the face-centred cubic lattice using square layers. As a bonus, we see that the Kepler conjecture is not quite as straightforward as greengrocers packing oranges.

Suppose that we start with a flat layer of spheres arranged in triangles, Figure 16 (right). In between the spheres are curved triangular dents, and a further layer of spheres can fit into these. When we began with a square layer, we were able to use all of the dents, so the position of the second layer, and those that followed, was uniquely determined. This is no longer the case if we begin with a triangular arrangement. We can't use all of the dents, because they are too close together. We can use only half of them. One choice is shown in Figure 18 (left), using small grey dots for clarity, and Figure 18 (right) shows how the next layer of spheres should be placed. The second way to fit a new layer into the dents of layer 1 is shown in Figure 19 (left) using darker dots. These dots coincide with dents in layer 2, so we add layer 3 in the corresponding positions: the result is Figure 19 (right).

The distinction between these choices doesn't actually make any difference when we have just these two layers. If we rotate the second arrangement through 60 degrees, we get the first. They are the same 'up to symmetry'. But after the first two layers have been positioned, there are two genuinely different choices for the third layer. Each new layer has two systems of dents, shown by the light and dark dots in Figure 19 (left). One system matches the centres of the layer immediately below, visible as small light grey triangles in Figure 19 (right). The other matches dents in the layer below that, visible as

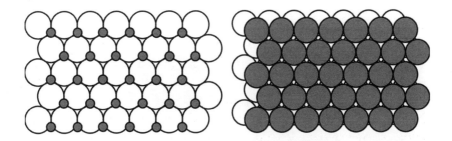

Fig 18 Fitting a triangular lattice into a set of gaps in the layer below.

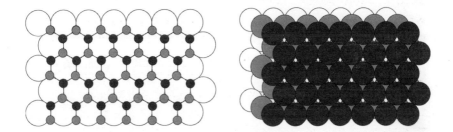

Fig 19 Stacking triangular lattices on top of each other.

triangles containing a tiny white hexagon in Figure 19 (right). To get the face-centred cubic lattice we must use the dark grey positions for the third layer, and then continue the same pattern indefinitely.

It's not entirely obvious that the result is the face-centred cubic lattice. Where are the squares? The answer is that they are present, but they slope at an angle. Figure 20 shows six successive triangular layers, with a number of spheres removed. The arrows indicate the rows and columns of a square lattice, hidden away inside. Layers parallel to this one are also square lattices, and they fit together in exactly the way I constructed the face-centred cubic lattice.

How 'tight' is this packing? We measure the tightness (efficiency, closeness) of a packing by its density: the proportion of space occupied by spheres.[40] The bigger the density, the tighter the pack. Cubes pack together with density 1, filling the whole of space. Spheres obviously have to leave gaps, so the density is less than 1. For the face-centred cubic lattice the density is exactly $\pi/\sqrt{18}$, which is roughly 0.7405. So for this packing, the spheres fill slightly less than three quarters of the

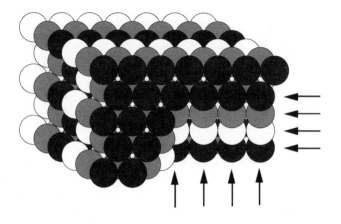

Fig 20 Hidden inside the triangular layers are square layers, at a slant.

space. The Kepler conjecture states that no sphere packing can have greater density than this.

I've stated that rather carefully. I've not said 'the face-centred cubic packing has greater density than any other'. That's false, spectacularly so. To see why, go back to the construction of the face-centred cubic lattice using triangular layers. I said that once the first two layers are determined, there are two choices for the third layer. The face-centred cubic lattice appears if we use the second one, the dark grey dots. What happens if we use the other one, the light grey dots? Now layer 3 sits exactly above layer 1. If we continue like that, placing each new layer immediately above the layer two stages below, we get a second lattice packing: the hexagonal lattice. It is genuinely different from the face-centred cubic packing, but it has the same density. This is intuitively clear because the two different ways to place the third layer are related by a rotational symmetry, so it fits the previous layer exactly as tightly, either way.

These are the only two *lattice* packings that can be obtained from successive triangular layers, but in 1883 the geologist and crystallographer William Barlow pointed out that we can choose the location of each successive layer at random from the two possibilities. Since either position makes the same contribution to the density, all of these packings have density $\pi/\sqrt{18}$. There are infinitely many random sequences, leading to infinitely many different packings, all having that density.

In short, there is no such thing as 'the' densest sphere packing. Instead, there are infinitely many of them, all equally dense. This lack of uniqueness is a warning: this is not a straightforward problem. The optimal *density* is unique, if Kepler was right, but infinitely many different arrangements have that density. So a proof that this density really is optimal is not just a matter of successively fitting each new sphere as tightly as possible. There are choices.

Impressive though the experience of greengrocers may be – and the face-centred cubic lattice was surely present in predynastic Egyptian markets – it's not remotely conclusive. In fact, it's a bit of an accident that the greengrocer's method yields a good answer. The problem facing greengrocers isn't that of packing oranges as closely as possible in space, where any arrangement is in principle possible. It is to pile up oranges stably, in a world where the ground is flat and gravity acts downwards. Greengrocers naturally start by making a layer; then they add another layer, and so on. They are likely to make the first layer into a square lattice if they're putting the oranges inside a rectangular box. If the oranges are unconfined then either a square or a triangular lattice is natural. As it happens, both give the same face-centred cubic lattice – at least, if the layers are appropriately placed in the triangular case. The square lattice actually looks like a poor choice, because it's not the densest way to pack a layer. By luck more than judgement, that turns out not to matter.

Physicists aren't interested in oranges. What they want to pack together are atoms. A crystal is a regular, spatially periodic arrangement of atoms. The Kepler conjecture explains the periodicity as a natural consequence of the atoms packing as closely as possible. As far as most physicists are concerned, the existence of crystals is evidence enough, so the conjecture is evidently true. However, we've just seen that there are infinitely many ways to pack spheres just as densely as the face-centred cubic and hexagonal lattices do, none of which are spatially periodic. So why does nature use periodic patterns for crystals? A possible answer is that we shouldn't model atoms as spheres.

Mathematicians aren't interested in oranges either. Like Kepler, they prefer to work with perfect, identical spheres. They don't find the physicists' argument convincing. If we shouldn't model atoms as spheres, the existence of crystals ceases to be evidence in favour of the

Kepler conjecture. We can't have it both ways. Even if you argue that the conjecture sort-of explains the crystal lattice, and the crystal lattice sort-of shows that the conjecture is correct ... there's a logical gap. Mathematicians want a proof.

Kepler didn't call his statement a conjecture: he just put it in his book. It's totally unclear whether he intended it to be interpreted in such a sweeping manner. Was he claiming that the face-centred cubic lattice was the 'closest pack in three dimensions' among all conceivable ways to pack spheres? Or did he just mean it was the closest pack of the three he had considered? We can't go back and ask. Whatever the historical reality, the interpretation of interest to mathematicians and physicists was the sweeping one, the ambitious one. The one that asked you to contemplate every possible way to pack infinitely many spheres together in infinite space – and show that none of them has greater density than the face-centred cubic lattice.

It is very easy to underestimate the difficulty of the Kepler conjecture. Surely the way to get the tightest pack is to add spheres one by one, making each one touch as many of the others as possible as you go? That leads inevitably to Kepler's pattern. And so it does if you add the spheres in the right order, placing them in the right positions when there are alternatives. However, there's no guarantee that this step-by-step process, adding spheres one at a time, can't be trumped by something more far-reaching. Anyone who has packed holiday gear into the boot of a car learns that fitting things in one by one can leave gaps into which nothing will fit, but starting again from the beginning and taking more care sometimes squeezes more in. Admittedly, part of the problem in packing holiday gear is the different shapes and sizes of the objects you're trying to fit in, but the logical point is clear enough: ensuring the tightest arrangement in a small region could have knock-on effects, and fail to lead to the tightest arrangement in a bigger region.

The arrangements that Kepler considers are very special. It is conceivable that some entirely different arrangement might be able to pack identical spheres together even more tightly. Maybe bumpy layers would be more efficient. Maybe 'layers' is the wrong idea. And even if you are absolutely certain that it's the right idea, you still have to prove that.

Not convinced? Still think it's obvious? So obvious that it doesn't *need* a proof? Let me try to destroy your confidence in your intuition for sphere packing. Here's a much simpler question, involving circles in the plane. Suppose I give you 49 identical circles, each of diameter 1 unit. What is the size of the *smallest* square that can contain them, if they are packed together without overlapping? Figure 21 (left) shows the obvious answer: pack them like milk bottles in a crate. The side of the crate is exactly 7 units. To prove this is the best, observe that each circle is held rigidly in place by all the others, so there's no way to create extra room. Figure 21 (right) shows that this answer is wrong. Pack them in the irregular manner shown, and they will fit into a square crate whose side is slightly less than 6.98 units.[41] So the proof is wrong as well. Being rigid is no guarantee that you can't do better.

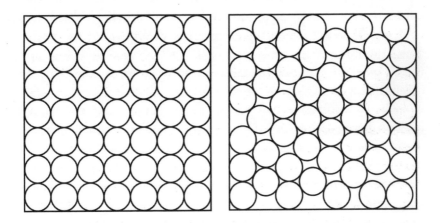

Fig 21 *Left*: 49 circles in a 7×7 square. *Right*: How to fit 49 circles into a slightly smaller square.

In fact, it's easy to see that the reasoning leading to the answer '7' can't possibly be right. Just consider larger squares. Using a square lattice, n^2 circles of diameter 1 pack into a square of side n. There is no way to improve the density by moving these circles in a continuous manner, because the packing is rigid. But there must be denser packings for large enough n, because a triangular lattice is more efficient than a square lattice. If we take a really big square, and fit as many circles into it as we can using a triangular lattice, the advantage that this has over the square lattice will eventually cause it to win, despite 'edge effects' at the boundary where you have to leave gaps. The size of the boundary is $4n$, which gets arbitrarily small compared to n^2. As it happens, the exact point at which the triangular lattice takes over is when $n = 7$. That's not obvious and it takes a lot of detailed work to establish it, but some n has to work. Rigidity is not enough.

There are really two versions of the Kepler conjecture. One considers only lattice packings, where the centres of the spheres form a spatially periodic pattern, repeating itself indefinitely in three independent directions like a kind of solid wallpaper. Even then the problem is hard, because there are many different lattices in space. Crystallographers recognise 14 types, classified by their symmetries, and some of those types are determined by numbers that can be adjusted to infinitely many different values. But the difficulties are compounded when we consider the second version of the problem, which permits all possible packings. Each sphere hovers in space, there's no gravity, and no obligation to form layers or other symmetric arrangements.

When a problem seems too hard, mathematicians put it on the back burner and look for simpler versions. Kepler's thoughts about flat layers of spheres suggest starting with circle packings in a plane. That is, given an infinite supply of identical circles, pack them together as closely as possible. Now the density is the proportion of *area* that the circles cover. In 1773 Joseph Louis Lagrange proved that the densest lattice packing of circles in a plane is the triangular lattice, with density $\pi/\sqrt{12} = 0.9069$. In 1831 Gauss was reviewing a book by Ludwig Seeber, who had generalised some of Gauss's number-theoretic results to equations in three variables. Gauss remarked that Seeber's results prove that the face-centred cubic and hexagonal lattices provide the densest lattice packing in three-dimensional space. An enormous amount is now known about lattice packings in spaces of higher dimension – 4, 5, 6, and so on. The 24-dimensional case is especially well understood. (The subject is like that.) Despite its air of impracticality, this area actually has implications for information theory and computer codes.

Non-lattice packings are another matter entirely. There are infinitely many, and they don't have any nice regular structure. So why not go to the other extreme and try random packings? In his *Vegetable Staticks* of 1727, Stephen Hales reported experiments in which he 'compressed several fresh parcels of pease [peas] in the same pot', finding that when they were all pressed together they formed 'pretty regular dodecahedrons'. He seems to have meant that the regular dodecahedrons were pretty, not that the dodecahedrons were pretty regular, but the second interpretation is better because regular dodecahedrons can't fill space. What he saw were probably rhombic dodecahedrons, which we've seen are associated with the face-centred cubic packing. G. David Scott put lots of ball bearings into a container, and shook it thoroughly, observing that the highest density was 0.6366. In 2008 Chaoming Song, Ping Wang, and Hernán Makse derived this figure analytically.[42] However, their result does not imply that Kepler was right – if only because, as stated, it would imply that the face-centred cubic lattice, with density 0.74, can't exist. The simplest way to explain this discrepancy is that their result ignores extremely rare exceptions. The face-centred cubic lattice, the hexagonal lattice, and all the arrangements of randomly chosen triangular layers are all exceptions of this kind. By the same token, there might exist some other arrangement with an even greater density. It can't be a lattice, but a random search will never find it because its probability is zero. So the study of random packings, while relevant to many questions in physics, doesn't tell us a lot about the Kepler conjecture.

The first real breakthrough came in 1892, when Axel Thue gave a lecture to the Scandinavian Natural Science Congress, sketching a proof that no circle packing in the plane can be denser than the triangular lattice. His lecture was published, but the details are too vague to reconstruct the proof he had in mind. He gave a new proof in 1910, which seemed convincing, save for a few technical points which he simply assumed could be sorted out. Instead of filling these gaps, László Fejes Tóth obtained a complete proof by other methods in 1940. Soon after, Beniamino Segre and Kurt Mahler found alternative proofs. In 2010 Hai-Chau Chang and Lih-Chung Wang put a simpler proof on the web.[43]

Finding the greatest density for circle or sphere packings, under specified conditions, falls into a general class of mathematical questions known as optimisation problems. Such a problem seeks the maximum or minimum value of some function, which is a mathematical rule for calculating a quantity that depends on some set of variables in a specific manner. The rule is often specified by a formula, but this is not essential. For instance, the milk crate problem, with 49 circles, can be formulated in this manner. The variables are the coordinates of the centres of the 49 circles; since each circle needs two coordinates, there are 98 variables. The function is the size of the smallest square, with sides parallel to the coordinate axes, that contains a given set of non-overlapping circles. The milk crate problem is equivalent to finding the minimum value that this function can attain as the variables range over all packings.

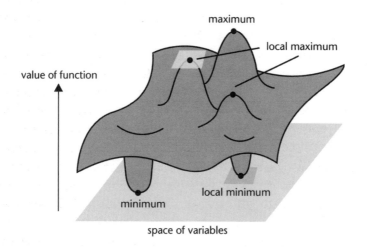

Fig 22 Peaks and valleys for a function.

A function can be thought of as a many-dimensional landscape. Each point in the landscape corresponds to a choice of the variables, and the height at that point is the corresponding value of the function. The maximum of the function is the height of the highest peak, and the minimum is the depth of the deepest valley. In principle, optimisation problems can be solved by calculus: the function must be horizontal at a peak or valley, Figure 22, and calculus expresses this condition as an equation. To solve the milk-crate problem by that method we would have to solve a system of 98 equations in the 98 variables.

One snag with optimisation problems is that equations like these often have a large number of solutions. A landscape can have lots of local peaks, only one of which is the highest. Think of the Himalayas: hardly anything *but* peaks, yet only Everest holds the altitude record. Methods for finding peaks, of which the most obvious is 'head uphill if you can', often get trapped in a local peak. Another snag is that as the number of variables grows, so does the likely number of local peaks. Nevertheless, this method sometimes works. Even partial results can be useful: if you find a local peak, the maximum must be at least that high. This is how the improved arrangement of circles in the milk-crate problem was found.

For lattice packings, the function whose maximum is sought depends only on finitely many variables, the directions and lengths along which the lattice repeats. For non-lattice packings, the function depends on infinitely many variables: the centres of all the circles or spheres. In such cases the direct use of calculus or other optimisation techniques is hopeless. Tóth's proof used a clever idea to reformulate the non-lattice packing problem for circles as an optimisation problem in a *finite* set of variables. Later, in 1953, he realised that the same trick could in principle be applied to the Kepler conjecture. Unfortunately, the resulting function depends on about 150 variables, far too many for hand calculation. But Tóth presciently saw a possible way out: 'Mindful of the rapid development of our computers, it is imaginable that the minimum may be determined with great exactitude.'

At that time, computing was in its infancy, and no sufficiently powerful machine existed. So subsequent progress on the Kepler conjecture followed different lines. Various mathematicians placed bounds – upper limits – on how dense a sphere packing could be. For example in 1958 Rogers proved that it is at most 0.7797: no rare exceptions, this bound applied to all sphere packings. In 1986 J.H. Lindsey improved the bound to 0.77844, and Douglas Muder shaved off a tiny bit more in 1988 to obtain a bound of 0.77836.[44] These results show that you can't do *much* better than the face-centred cubic lattice's 0.7405. But there was still a gap, and little prospect of getting rid of it.

In 1990 Wu-Yi Hsiang, an American mathematician, announced a proof of the Kepler conjecture. When the details were made public, however, doubts quickly set in. When Tóth reviewed the paper in *Mathematical Reviews*, he wrote: 'If I am asked [whether the paper provides] a proof of the Kepler conjecture, my answer is: no. I hope that Hsiang will fill in the details, but I feel that the greater part of the work has yet to be done.'

Thomas Hales, who had been working on the conjecture for many years, also doubted that Hsiang's method could be repaired. Instead, he decided it was time to take Tóth's approach seriously. A new generation of mathematicians had grown up, for whom reaching for a computer was more natural than reaching for a table of logarithms. In 1996 Hales outlined a proof strategy based on Tóth's idea. It required identifying all possible ways to arrange several spheres in the immediate vicinity of a given one. A sphere packing is determined by the centres of the spheres; for unit spheres, these must be at least 2 units apart. Say that two spheres are *neighbours* if their centres are at most 2.51 units apart. This value is a matter of judgement: make it too small and there isn't enough room to rearrange neighbours to improve the density; make it too big and the number of ways to arrange the neighbours becomes gigantic. Hales found that 2.51 was an effective compromise. Now we can represent how neighbours are arranged by forming an infinite network in space. Its dots are the centres of the spheres, and two dots are joined by a line if they are neighbours. This network is a kind of skeleton of the packing, and it contains vital information about the neighbourhood of each sphere.

For any given sphere, we can look at its immediate neighbours in the network and consider only the lines between these neighbours, omitting the original sphere. The result is a sort of cage surrounding the dot at the centre of the original sphere. Figure 23 (left pair) shows the neighbours of a sphere in the face-centred cubic lattice and the associated cage. Figure 23 (right pair) does the same for a special arrangement of spheres, the pentagonal prism, which turned out to be a key player in the proof. Here there are two bands of pentagons parallel to the 'equator' of the central sphere, plus a single sphere at each pole.

The cages form a solid with flat faces, and the geometry of this solid controls the packing density near the central sphere.[45] The key

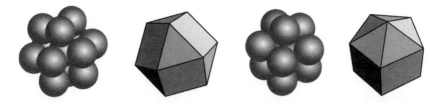

Fig 23 *From left to right:* Neighbourhood of a sphere in the face-centred cubic lattice; the cage formed by its neighbours; neighbourhood of a sphere of pentagonal prism type; the cage formed by its neighbours.

idea is to associate to each cage a number, known as its score, which can be thought of as a way to estimate the density with which the sphere's neighbours are packed. The score isn't the density itself, but a quantity that is better behaved and easier to calculate. In particular, you can find the score of the cage by adding up scores related to its faces, which doesn't work for density. In general many different notions of score satisfy that condition, but they all agree on one thing: for the face-centred cubic and hexagonal lattices the score is always 8 'points', no matter what choices are made in its definition. Here a point is a specific number:

$$4 \arctan \frac{\sqrt{2}}{5} - \frac{\pi}{3} = 0.0553736$$

So 8 points is actually 0.4429888. This curious number comes from the special geometry of the face-centred cubic lattice. Hales's key observation relates the Kepler conjecture to this number: if every cage has a score of 8 points or less, then the Kepler conjecture is true. So the focus shifts to cages and scores.

Cages can be classified by their topology: how many faces they have with a given number of sides, and how those faces adjoin. For a given topology, however, the edges can have many different lengths. These lengths affect the score, but the topology lumps lots of different cages together, and these can be dealt with in the same general way. In his eventual proof, Hales considered about 5000 types of cage, but the main calculations concentrated on a few hundred. In 1992, he proposed a five-stage programme:

1 Prove the desired result when all faces of the cage are triangles.

2   Show that the face-centred cubic and hexagonal packings have a higher score than any cage with the same topology.

3   Deal with the case when all faces of the cage are triangles and quadrilaterals, with the exception of the pentagonal prism, which is harder.

4   Deal with any cage having a face with more than four edges.

5   Sort out the only case remaining, when the cage is a pentagonal prism.

Part 1 was solved in 1994, and part 2 in 1995. As the programme developed, Hales modified the definition of a cage to simplify the argument (his term is 'decomposition star'). The new definition doesn't alter the two cages illustrated, and it didn't have any serious effects on those parts of the proof that had already been obtained. By 1998, using this new concept, all five stages had been completed. Hales's student Samuel Ferguson solved Part 5, the tricky case of a pentagonal prism.

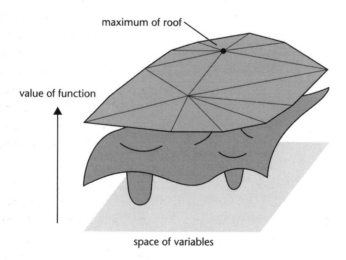

Fig 24 Fitting a roof over the top of a function.

The analysis involved heavy use of a computer at all stages. The trick is to pick, for each local network, a notion of score that makes the calculation relatively easy. Geometrically, replacing the density by the score puts a kind of roof over the top of the smooth landscape whose peak is being sought. The roof is made from numerous flat pieces, Figure 24. Shapes like this are easier to work with than smooth surfaces, because the maxima must occur at the corners, and these can be found by solving much simpler equations. There are efficient methods for doing this, known as linear programming. If the roof has cunningly been constructed so that its peak coincides with the peak of the smooth surface, then this simpler computation locates the peak of the smooth surface.

There is a price to pay for this approach: you have to solve about 100,000 linear programming problems. The calculations are lengthy, but well within the capabilities of today's computers. When Hales and Ferguson prepared their work for publication it ran to about 250 pages of mathematics, plus 3 gigabytes of computer files.

In 1999 Hales submitted the proof to the *Annals of Mathematics*, and the journal chose a panel of 12 expert referees. By 2003, the panel declared itself '99 per cent certain' that the proof was correct. The remaining uncertainty concerned the computer calculations; the panel had repeated many of them, and otherwise checked the way the proof was organised and programmed, but they were unable to verify some aspects. After a delay, the journal published the paper. Hales recognised that this approach to the proof would probably never be certified 100 per cent correct, so in 2003 he announced that he was starting a project to recast the proof in a form that could be verified by a computer using standard automated proof-checking software.

This may sound like stepping out of the frying-pan to fall into the fire, but it actually makes excellent sense. The proofs that mathematicians publish in journals are intended to convince humans. As I said in chapter 1, that sort of proof is a kind of story. Computers are poor at telling stories, but they excel at something we are hopeless at: performing long, tedious calculations without making a mistake. Computers are ideal for the formal concept of proof in the undergraduate textbooks: a series of logical steps, each following from previous ones.

Computer scientists have exploited this ability. To check a proof, get a computer to verify each logical step. It ought to be easy, but the proofs in the journals aren't written like that. They leave out anything routine or obvious. The traditional phrases are easy to spot: 'It is easy to verify that …' 'Using the methods of Cheesberger and Fries, modified to take account of isolated singularities, we see that …' 'A short calculation establishes …' Computers can't (yet) handle that kind of thing. But humans can rewrite proofs with all these gaps filled in, and computers can then verify each step.

The reason that we're not hopping straight back into fire territory is straightforward: the software that does the verification needs to be checked only *once*. It is general-purpose software, applicable to all proofs written in the right format. All the worries about computer proofs are concentrated in that one piece of software. Verify that, and it can be used to verify everything else. You can even 'bootstrap' the process by writing the proof verification software in a language that can be checked by a much simpler piece of proof verification software.

In recent years, proofs of many key mathematical theorems have been verified in this manner. Often the proofs have to be presented in a style that is more suitable for computer manipulation. One of the current triumphs is a verified proof of the Jordan curve theorem: every closed curve in the plane that does not cross itself divides the plane into two distinct connected regions. This may sound obvious, but the pioneers of topology had trouble finding a rigorous proof. Camille Jordan finally succeeded in 1887 with a proof over 80 pages long, but he was later criticised for making unsubstantiated assumptions. Instead, the credit went to Oswald Veblen, who gave a more detailed proof in 1905, saying '[Jordan's] proof … is unsatisfactory to many mathematicians. It assumes the theorem without proof in the important special case of a simple polygon, and of the argument from that point on, one must admit at least that all details are not given.' Later mathematicians accepted Veblen's criticism without demur, but recently Hales went over Jordan's proof and found 'nothing objectionable' in it. In fact, Veblen's remark about a polygon is bizarre: the theorem is straightforward for a polygon, and Jordan's proof doesn't rely on that version anyway.[46] Storytelling proofs have their own dangers. It's always worth checking whether the popular version of the story is the same as the original one.

As a warm-up to the Kepler conjecture, Hales gave a formal computer-verified proof of the Jordan curve theorem in 2007, using

60,000 lines of computer code. Soon after, a team of mathematicians produced another formal proof using different software. Computer verification isn't totally foolproof, but neither are traditional proofs. In fact, many mathematical research papers probably contain a technical error somewhere. These errors show up from time to time, and mostly they turn out to be harmless. Serious errors are usually spotted because they introduce inconsistencies, so that something visibly makes no sense. This is another downside of the storytelling approach: the price we pay for making a proof comprehensible by humans is that a ripping yarn can sometimes be very convincing even when it's wrong.

Hales calls his approach *Project FlysPecK* – the F, P, and K standing for 'formal proof of Kepler'. Initially, he estimated that it would take about 20 years to complete the task.[47] Nine years into the project, considerable progress has already been made. It may finish early.

6

New solutions for old
Mordell Conjecture

NOW WE'RE HEADING BACK INTO the realms of number theory, aiming towards Fermat's last theorem. To prepare the ground I'll start with a less familiar but arguably even more important problem. In 2002 Andrew Granville and Thomas Tucker introduced it like this:[48]

> In [1922] Mordell wrote one of the greatest papers in the history of mathematics... At the very end of the paper, Mordell asked five questions which were instrumental in motivating much of the important research in Diophantine arithmetic in the twentieth century. The most important and difficult of these questions was answered by Faltings in 1983 by inventing some of the deepest and most powerful ideas in the history of mathematics.

Mordell is the British number theorist Louis Mordell, who was born in the United States to a Jewish family of Lithuanian origin, and Faltings is the German mathematician Gerd Faltings. The question referred to became known as the Mordell conjecture, and the quote gives away its current status: proved, brilliantly, by Faltings.

The Mordell conjecture belongs to a major area of number theory: Diophantine equations. They are named after Diophantus of Alexandria, who wrote a famous book, the *Arithmetica* ('Arithmetic') around AD 250. It is thought that originally the *Arithmetica* included 13 books, but only six have survived, all later copies. This was not an arithmetic text in the sense of addition and multiplication sums. It was the first algebra text, and it collected most of what the Greeks knew about how to solve equations. It even had a rudimentary form of algebraic notation, which is believed to have used a variant ς of the Greek letter sigma for the unknown (our x), Δ^Y for its square (our x^2), and K^Y for its cube (our x^3). Addition was denoted by placing symbols next to each other, subtraction had its own special symbol, the reciprocal of the unknown (our $1/x$) was ς^χ, and so on. The symbols have been reconstructed from later copies and translations, and may not be entirely accurate.

In the spirit of classical Greek mathematics, the solutions of equations sought in the *Arithmetica* were required to be rational numbers – that is, fractions like 22/7 formed from whole numbers. Often they were required to *be* whole numbers. All numbers involved were positive: negative numbers were introduced several centuries later in China and India. We now call such problems Diophantine equations. The book includes some remarkably deep results. In particular Diophantus appears to be aware that every whole number can be expressed as the sum of four perfect squares (including zero). Lagrange gave the first proof in 1770. The result that interests us here is a formula for all Pythagorean triples, in which two perfect squares add to form another perfect square. The name comes from Pythagoras's theorem: this relation holds for the sides of a right-angled triangle. The best known example is the celebrated 3-4-5 triangle: $3^2 + 4^2 = 5^2$. Another is $5^2 + 12^2 = 13^2$. There are infinitely many of these Pythagorean triples, and there is a recipe for finding them all in two lemmas (auxiliary propositions) preceding Propositions 29 and 30 in Book X of Euclid's *Elements*.

Euclid's procedure yields infinitely many Pythagorean triples. Mordell knew several other Diophantine equations for which there exists a formula that yields infinitely many solutions. He also knew another type of Diophantine equation with infinitely many solutions, *not* prescribed by a formula. These are the so-called elliptic curves – a rather silly name since they have virtually nothing to do with ellipses – and the infinitude of solutions arises because any two solutions can be combined to give another one. Mordell himself proved one of the basic properties of these equations: you need only a finite number of solutions to generate them all through this process.

Aside from these two types of equation, every other Diophantine equation that Mordell could think of fell into one of two categories.

Either it was known to have only finitely many solutions, including none at all, or no one knew whether the number of solutions was finite or infinite. This alone wasn't news, but Mordell thought he could spot a pattern that no one else had noticed. It wasn't a number-theoretic pattern at all; it came from topology. What mattered was how many holes the equation had. And to make sense of that, you had to think about its solutions in complex numbers, not rational numbers or integers. Which somehow seemed contrary to the entire spirit of Diophantine equations.

It's worth filling in a few details here. They will help a lot later. Don't be put off by the algebra; it's mainly there to give me something specific to refer to. Concentrate on the story behind the algebra.

Pythagorean triples are solutions, in integers, of the Pythagorean equation

$$x^2 + y^2 = z^2$$

Dividing by z^2 gives

$$(x/z)^2 + (y/z)^2 = 1$$

By chapter 3, this tells us that the pair of rational numbers $(x/z, y/z)$ lies on the unit circle in the plane. Now, the Pythagorean equation originated in geometry, and its interpretation is that the associated triangle has a right angle. The formula I've just derived provides a slightly different geometric interpretation, not just of one Pythagorean triple, but of all of them. The solutions of the Pythagorean equation correspond directly and naturally to all of the rational points on the unit circle. Here a point is said to be rational provided both of its coordinates are.

You can deduce a lot of interesting facts from this connection. With a bit of trigonometry, or by direct algebra, you can discover that for any number t the point

$$\left(\frac{2t}{t^2 + 1}, \frac{t^2 - 1}{t^2 + 1} \right)$$

lies on the unit circle. Moreover, if t is rational, so is this point. All rational points arise in this manner, so we have a complete formula for all solutions of the Pythagorean equation. It is equivalent to Euclid's formula, which is the same as Diophantus's. As an example, if $t = 22/7$ then the formula yields

$$\left(\frac{308}{533}, \frac{435}{533}\right)$$

and you can check that $308^2 + 435^2 = 533^2$. For us, the precise formula is not terribly important; what matters is that there is one.

This is not the only Diophantine equation for which a formula gives all solutions, but they are relatively rare. Others include the so-called Pell equations, such as $x^2 = 2y^2 + 1$. This has infinitely many solutions, such as $3^2 = 2 \times 2^2 + 1$, $17^2 = 2 \times 12^2 + 1$, and there is a general formula. However, Pythagorean triples have more structure than that, also derived from the geometry. Suppose you have two Pythagorean triples. Then there are two corresponding solutions of the Pythagorean equation – rational points on the circle. Geometry provides a natural way to 'add' those points together. Start from the point (1,0) at which the circle cuts the horizontal axis, and find the angles between this point and the two solutions. Add the two angles together, Figure 25, and see what point results. It certainly lies on the circle. A short calculation shows that it is rational. So from any two solutions, we can derive a third. Mathematicians had already noticed many facts like this. Most of them make immediate sense if you think of rational points on a circle.

The 'short calculation' that I slid over makes use of trigonometry. The classical trigonometric functions such as the sine and cosine are intimately related to the geometry of a circle. The calculation alluded to uses standard, rather elegant formulas for the sine and cosine of the sum of two angles in terms of the sines and cosines of the angles themselves. There are many ways to set up sines and cosines, and a rather neat one comes from integral calculus. If you integrate the algebraic function $1/\sqrt{1 - x^2}$, the result can be expressed in terms of the sine function. In fact, what we need is the inverse function of the sine: the angle whose sine is the number we're thinking of.[49]

The integral arises when we try to derive a formula for the arc length of a circle using calculus, and the geometry of the circle has a

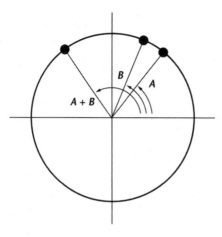

Fig 25 Combining two rational solutions *A* and *B* of the Pythagorean equation to obtain a third, $A + B$.

simple but very important implication for the result. The circumference of the unit circle is 2π, so going round the circle a distance 2π brings you back to the exact same point. The same goes for any integer multiple of 2π: by the standard mathematical convention, positive integers correspond to the anticlockwise direction, negative ones to the clockwise direction. It follows that the sine and cosine of a number remain unchanged if an integer multiple of 2π is added to that number. We say that the functions are periodic, with period 2π.

The analysts of the eighteenth and nineteenth centuries discovered a vast generalisation of this integral, along with a host of interesting new functions analogous to the familiar trigonometric ones. These new functions were intriguing; they were periodic, like the sine and cosine, but in a more elaborate way. Instead of having one period, like 2π (and its integer multiples), they had two independent periods. If you try to do that with real functions, all you get are constants, but for complex functions, the possibilities are much richer.

The area was initiated by the Italian mathematician Giulio di Fagnano and the prolific Euler. Fagnano was trying to find the arc length of an ellipse using calculus, but he couldn't find an explicit formula – no longer a surprise since we now know there isn't one.

However, he did notice a relationship between the lengths of various special arcs, which he published in 1750. Euler noticed the same relationship in the same context, and presented it as a formal relation between integrals. They are similar to the one associated with the sine function, but the quadratic expression $1 - x^2$ under the square root is replaced by a cubic or quartic polynomial, for example the quartic $(1 - x^2)(1 - 4x^2)$.

In 1811 Adrien-Marie Legendre published the first book in a massive three-volume treatise on these integrals, which are known as elliptic integrals because of their connection with the arc length of a segment of an ellipse. He managed to overlook the most significant feature of these integrals, however: the existence of new functions, analogous to the sine and cosine, whose *inverse* functions express the value of the integral in a simple way.[50] Gauss, Niels Henrik Abel, and Carl Jacobi quickly spotted the oversight. Gauss, rather typically, kept the discovery to himself. Abel submitted a paper to the French Academy in 1826, but Cauchy, the president, mislaid the manuscript, and it was not published until 1841, twelve years after Abel's tragic early death from lung disease. However, another paper by Abel on the same topic was published in 1827. Jacobi made these new 'elliptic functions' the basis of a huge tome, published in 1829, which propelled complex analysis along an entirely new trajectory.

What emerged was a beautiful package of interrelated properties, analogous to those of the trigonometric functions. The relationship noticed by Fagnano and Euler could be reinterpreted as a simple list of formulas relating elliptic functions of the sum of two numbers to elliptic functions of the numbers themselves. The most wonderful feature of elliptic functions outdoes trigonometric functions in a spectacular way. Not only are elliptic functions periodic: they are doubly periodic. A line is one-dimensional, so patterns can repeat in only one direction, along the line. The complex plane is two-dimensional, so patterns can repeat like wallpaper: down the strip of paper, and also sideways along the wall into adjacent strips of paper. Associated with each elliptic function are two independent complex numbers, its periods, and adding either of them to the variable does not change the value of the function.

Repeating this process, we conclude that the value of the function does not change if we add any integer combination of the two periods to the variable. These combinations have a geometric interpretation: they determine a lattice in the complex plane. The lattice specifies a tiling of the plane by parallelograms, and whatever happens in one parallelogram is copied in all the others, Figure 26. If we consider just one parallelogram, the way it joins up with adjacent copies means that we have to identify opposite edges, in the same way that a torus is defined by identifying opposite edges of a square, Figure 12. A parallelogram with opposite edges identified is also a topological torus. So, just as the sine and cosine are related to the circle, elliptic functions are related to a torus.

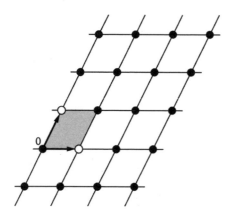

Fig 26 Lattice in the complex plane. The arrows point to the two periods, shown as white dots. The value of the function in the shaded parallelogram determines it in every other parallelogram.

There is also a link to number theory. I said that the inverse sine function is obtained by integrating a formula involving the square root of a quadratic polynomial. Elliptic functions are similar, but the quadratic is replaced by a cubic or quartic polynomial. The quartic case was mentioned briefly above, because historically it came first, but now let's focus on the cubic case. If we denote the square root by y, and the polynomial by $ax^3 + bx^2 + cx + d$ where a, b, c, d are numerical coefficients, then x and y satisfy the equation

$$y^2 = ax^3 + bx^2 + cx + d$$

This equation can be thought of in several different contexts, depending on what restrictions are placed on the variables and coefficients. If they are real, the equation defines a curve in the plane. If they are complex, algebraic geometers still call the set of solutions a curve, by analogy. But now it is a curve in the space of pairs of complex numbers, which is four-dimensional in real coordinates. And the curve is actually a surface, from this real-number viewpoint.

Figure 27 shows the real elliptic curves $y^2 = 4x^3 - 3x + 2$ and $y^2 = 4x^3 - 3x$, which are typical. Because y appears as its square, the curve is symmetric about the horizontal axis. Depending on the coefficients, it is either a single sinuous curve or it has a separate oval component. Over the complex numbers, the curve is always one connected piece.

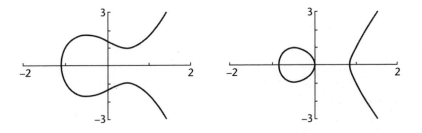

Fig 27 Typical real elliptic curves. *Left: $y^2 = 4x^3 - 3x + 2$. Right: $y^2 = 4x^3 - 3x$.*

Number theory comes into play when we require the variables and coefficients to be rational. Now we are looking at a Diophantine equation. It is rather confusingly called an elliptic curve, even though it looks nothing like an ellipse, because of the link to elliptic functions. It's like calling a circle a triangular curve because of the link to trigonometry. Unfortunately, the name is now inscribed in tablets of stone, so we have to live with it.

Because elliptic functions have a deep and rich theory, number theorists have discovered innumerable beautiful properties of elliptic curves. One is closely analogous to the way we can combine two solutions of the Pythagorean equation by adding the associated angles. Two points on an elliptic curve can be combined by drawing a straight line through them and seeing where it meets the curve for a third time, Figure 28. (There always is such a third point, because the equation is a cubic. However, it might be 'at infinity', or it might coincide with one of the first two points if the line cuts the curve at a tangent.) If the two points are P and Q, denote the third one by P * Q.

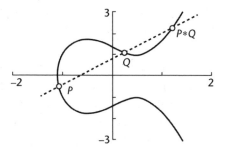

Fig 28 Combining points P, Q to get the point P * Q.

A calculation shows that if P and Q are rational points then so is P * Q. The operation * gives the set of rational points an algebraic structure, but it turns out to be useful to consider a related operation. Choose any rational point O on the curve, and define

$$P + Q = (P * Q) * O$$

This new operation obeys some basic laws of ordinary algebra, with O behaving like zero, and it turns the set of all rational points into what algebraists call a group, chapter 10. The essential point is that like Pythagorean triples, you can 'add' any two solutions to get a third. The occurrence of this 'group law' on the rational points is striking, and in particular it means that once we have found two rational solutions of the Diophantine equation, we automatically get many more.

Around 1908 Poincaré asked whether there exists a finite number of solutions from which all other solutions can be obtained by applying the group operation over and over again. This result is important because it implies that *all* rational solutions can be characterised by writing down a finite list. In his spectacular 1922 paper Mordell proved that the answer to Poincaré's question is 'yes'. Now elliptic curves became of central importance in number theory, because it was unusual to have that kind of control over any Diophantine equation.

Both the Pythagorean equation and elliptic curves, then, have infinitely many rational solutions. In contrast, many Diophantine equations have only finitely many solutions, often none at all. I'm going to digress slightly to discuss an entire family of such equations, and the recent remarkable proof that the obvious solutions are the only ones that exist.

The Pythagoreans were interested in their equation because they believed that the universe is founded on numbers. In support of this philosophy, they discovered that simple numerical ratios govern musical harmony. They observed this experimentally using a stretched string. A string of the same tension that is half as long plays a note one octave higher. This is the most harmonious combination of two notes: so harmonious that it sounds a bit bland. In Western music the next most important harmonies are the fourth, where one string is $\frac{3}{4}$ as long as the other, and the fifth, where one string is $\frac{2}{3}$ as long as the other.[51]

Starting with 1 and repeatedly multiplying by 2 or 3 yields numbers 2, 3, 4, 6, 8, 9, 12, and so on – numbers of the form $2^a 3^b$. Because of the musical connection these became known as harmonic numbers. In the thirteenth century a Jewish writer who lived in France wrote *Sha'ar ha-Shamayim* ('Door of heaven'), an encyclopaedia based on Arab and Greek sources. He divided it into three parts: physics, astronomy, and metaphysics. His name was Gerson ben Solomon Catalan. In 1343 the Bishop of Meaux persuaded Gerson's son (well, historians think it was probably his son) Levi ben Gerson to write a mathematical book, *The Harmony of Numbers*. It included a problem raised by the composer and musical theorist Philippe de Vitry: when can two harmonic numbers differ by 1? It is easy to find such pairs: de Vitry knew of four, namely (1, 2), (2, 3), (3, 4), and (8, 9). Ben Gerson proved that these are the only possible solutions.

Among de Vitry's pairs of harmonic numbers, the most interesting is (8, 9). The first is a cube, 2^3; the second is a square, 3^2. Mathematicians started to wonder whether other squares and cubes might differ by 1, and Euler proved that they could not, aside from the trivial case (0, 1), and also (−1, 0) if negative integers are allowed. In 1844 the second Catalan in the story committed to print a more sweeping claim, which many mathematicians must have thought of but hadn't bothered to make explicit. He was the Belgian mathematician

Eugène Charles Catalan, and in 1844 he wrote to one of the leading mathematics journals of the day, the *Journal für die Reine und Angewandte Mathematik* ('Journal for pure and applied mathematics'):

> I beg you, sir, to please announce in your journal the following theorem that I believe true although I have not yet succeeded in completely proving it; perhaps others will be more successful. Two consecutive whole numbers, other than 8 and 9, cannot be consecutive powers; otherwise said, the equation $x^m - y^n = 1$ in which the unknowns are positive integers only admits a single solution.

This statement became known as the Catalan conjecture. The exponents m and n are integers greater than 1.

Despite partial progress, the Catalan conjecture stubbornly refused to surrender, until it was dramatically proved in 2002 by Preda Mihăilescu. Born in Romania in 1955, he had settled in Switzerland in 1973, and had only recently completed his PhD. The title of his thesis was 'Cyclotomy of rings and primality testing', and it applied number theory to primality testing, chapter 2. This problem had no particular bearing on the Catalan conjecture, but Mihăilescu came to realise that his methods most certainly did. They derived from ideas that I mentioned in chapter 3: Gauss's construction of the regular 17-gon and associated algebraic equations, whose solutions are called cyclotomic numbers. The proof was highly technical and came as a shock to the mathematical community. It tells us that whichever values we choose for the two powers, the number of solutions is finite – and aside from obvious solutions using 0 and ± 1, the only interesting one is $3^2 - 2^3 = 1$.

The above examples show that some Diophantine equations have infinitely many solutions, others don't. Big deal – those alternatives cover everything. If you start asking which equations are of which type, however, it gets more interesting. Mordell, an expert on Diophantine equations, was writing a seminal textbook. In his day the area looked like early biology: a lot of butterfly collecting and very little in the way of systematic classification. Here was a Painted Pythagorean, there a Large Blue Elliptic, and in the bushes were caterpillars of the Speckled

Pellian. The field was just as Diophantus had left it, only more so: a structureless list of separate tricks, one for each type of equation. This is poor textbook material, and it desperately needed organising, so Mordell set out to do just that.

At some point he must have noticed that all of the equations known to have infinitely many rational solutions – such as the Pythagorean equation and elliptic curves – had a common feature. He focused on one class of equations, those that (after being converted to rational number equations, as I did for Pythagoras) involve just two variables. There are two cases where we know how to find infinitely many solutions. One is exemplified by the Pythagorean equation in the equivalent form $x^2 + y^2 = 1$. Here there is a formula for the solutions. Plug any rational number into the formula and you get a rational solution, and all solutions arise. The other is exemplified by elliptic curves: there is a *process* that generates new solutions from old ones, and a guarantee that if you start with a suitable finite set of solutions, this process produces them all.

Mordell's conjecture states that whenever there are infinitely many rational solutions, one of these two features must apply. Either there is a general formula, or there is a process that generates all solutions from a suitable finite set of them. In all other cases, the number of rational solutions is finite, for example the equations $x^m - y^n = 1$ that feature in the Catalan conjecture. In a sense, the solutions are then just coincidences, with no underlying structure.

Mordell arrived at this observation in a slightly different way. He noticed that every equation with infinitely many rational solutions has a striking topological feature. It has genus 0 or 1. Recall from chapter 4 that genus is a concept from the topology of surfaces, and it counts how many holes the surface has. A sphere has genus 0, a torus genus 1, a 2-holed torus has genus 2, and so on. How do surfaces come into a problem in number theory? From coordinate geometry. We saw that the Pythagorean equation, interpreted in terms of rational numbers and extended to allow real numbers as solutions, determines a circle. Mordell went one step further, and allowed complex numbers as solutions. Any equation in two complex variables determines what algebraic geometers call a complex curve. However, from the point of view of real numbers and the human visual system, every complex number is two-dimensional: it has two real components, its real and imaginary parts. So a 'curve' to complex eyes is a surface to you and me. Being a surface, it has a genus: there you go.

Whenever an equation was known to have only finitely many solutions, its genus was at least 2. Important equations whose status was unknown also had genus at least 2. In a wild and courageous leap, based on what seemed at the time to be pretty flimsy evidence, Mordell conjectured that any Diophantine equation with genus 2 or more has only finitely many rational solutions. At a stroke, the Diophantine butterflies were neatly arranged into related families; appropriately, by genus.

There was only one tiny snag with Mordell's conjecture. It related two extremely different things: rational solutions and topology. At the time, any plausible link was tenuous in the extreme. If a connection existed, no one knew how to go about finding it. So the conjecture was wild, unsubstantiated speculation, but the potential payoff was huge.

In 1983 Faltings published a dramatic proof that Mordell's wild speculation had, in fact, been right. His proof used deep methods from algebraic geometry. A very different proof, based on approximating real numbers by rational ones, was soon found by Paul Vojta, and Enrico Bombieri published a simplified proof along the same lines in 1990. There is an application of Faltings's theorem to Fermat's last theorem, a problem that we treat at length in chapter 7. This states that for any integer n greater than or equal to 3, the equation $x^n + y^n = 1$ has only finitely many integer solutions. The genus of the associated curve is $(n - 1)(n - 2)/2$, and this is at least 3 if n is 4 or greater. Faltings's theorem immediately implies that for any $n \geqslant 4$, the Fermat equation has at most a finite number of rational solutions. Fermat claimed it had none except when x or y is zero, so this was a big advance. In the next chapter we take up the story of Fermat's last theorem, and see how Fermat's claim was vindicated in full.

7

Inadequate margins
Fermat's Last Theorem

WE FIRST ENCOUNTERED FERMAT in chapter 2, where his elegant theorem about powers of numbers provided a method for testing numbers to see whether they are prime. This chapter is about a far more difficult assertion: Fermat's last theorem. It sounds very mysterious. 'Theorem' seems clear, but who was Fermat, and why was it his *last* theorem? Is the name a cunning marketing ploy? As it happens, no: the name became attached to the problem in the eighteenth century, when only a few leading mathematicians had heard of it or cared about it. But Fermat's last theorem really is mysterious.

Pierre Fermat was born in France in 1601, according to some sources, and in 1607–8, according to others. The discrepancy may perhaps come from confusion with a sibling of the same name. His father was a successful leather merchant and held a high position in local government, and his mother came from a family of parliamentary lawyers. He went to the University of Toulouse, moved to Bordeaux in the late 1620s, and there he showed promising signs of mathematical talent. He was fluent in several languages, and he put together a restoration of a lost work of classical Greek mathematics by Apollonius. He shared his many discoveries with leading mathematicians of the day.

In 1631, having taken a law degree at the University of Orléans, he was appointed a councillor of the High Court of Judicature in Toulouse. This entitled him to change his surname to 'de Fermat', and he remained a councillor for the rest of his life. His passion, though, was mathematics. He published little, preferring to write letters outlining his discoveries, usually without proof. His work received due recognition from the professionals, with many of whom he was on first-name terms, while retaining his amateur status. But Fermat was so talented that he was really a professional; he just didn't hold an official position in mathematics.

Some of his proofs have survived in letters and papers, and it is clear that Fermat knew what a genuine proof was. After his death, many of his deepest theorems remained unproved, and the professionals set to work on them. Within a few decades, only one of Fermat's statements still lacked a proof, so naturally, this became known as his last theorem. Unlike the others, it failed to succumb, and quickly became notorious for the contrast between its ease of statement and the apparent difficulty of finding a proof.

Fermat seems to have conjectured his last theorem around 1630. The exact date is not known, but that was when Fermat started reading a recently published edition of Diophantus's *Arithmetica*. And that's where he got the idea. The last theorem first saw print in 1670, five years after Fermat's death, when his son Samuel published an edition of the *Arithmetica*. This edition had a novel feature. It incorporated marginal notes that Pierre had written in the margins of his personal copy of the 1621 Latin translation by Claude Gaspard Bachet de Méziriac. The last theorem is stated as a note attached to Diophantus's question VIII of Book II, Figure 29.

The problem solved there is to write a perfect square as the sum of two perfect squares. In chapter 6 we saw that there are infinitely many of these Pythagorean triples. Diophantus asks a related but harder question: how to find the two smaller sides of the triangle, given the largest. A specific square must be 'divided' into two squares, that is, expressed as their sum. He shows how to solve this problem when the largest side of the triangle is 4, obtaining the answer

$$4^2 = (16/5)^2 + (12/5)^2$$

in rational numbers. Multiplying through by 25 we obtain $20^2 = 16^2 + 12^2$, and dividing by 16 yields the familiar $3^2 + 4^2 = 5^2$. Diophantus typically illustrated general methods with specific

QVÆSTIO VIII.

PRoposum quadratum diuidere in duos quadratos. Imperatum fit vt 16. diuidatur in duos quadratos. Ponatur primus 1 Q.Oportet igitur 16 — 1 Q.æquales effe quadrato. Fingo quadratum à numeris quotquot libuerit, cum defectu tot vnitatum quod continet latus ipfius 16. efto à 2 N. — 4. ipfe igitur quadratus erit, 4 Q. + 16. — 16 N. hæc æquabuntur vnitatibus 16 — 1 Q. Communis adiiciatur vtrimque defectus, & à fimilibus auferantur fimilia, fient 5 Q. æquales 16 N. & fit 1 N. ⁴⁄₅ Erit igitur alter quadratorum ²⁵⁶⁄₂₅. alter verò ¹⁴⁴⁄₂₅ & vtriufque fumma eft ⁴⁰⁰⁄₂₅ feu 16. & vterque quadratus eft.

ΤΟΝ ἐπιταχθέντα τετράγωνον διελεῖν εἰς δύο τετραγώνους. ἐπιτετάχθω δὴ ὁ ιϛ διελεῖν εἰς δύο τετραγώνους. καὶ τετάχθω ὁ πρῶτος δυνάμεως μιᾶς. δεήσει ἄρα μονάδας ιϛ λείψει δυνάμεως μιᾶς ἴσας ᾗ τετραγώνῳ. πλάσω τὸν τετράγωνον ἀπὸ ςϛ. ὅσων δὴ ποτε λείψει τοσούτων μὸ ὅσων ἐστὶν ἡ τ̄ ιϛ, μὸ πλήθως. ἴσω ςϛ β λείψει μὸ δ. αὐτὸς ἄρα ὁ τετράγωνος ἴσαι δυνάμεων δ μὸ ιϛ λείψει ςϛ ιϛ. ταῦτα ἴσα μονάσι ιϛ λείψει δυνάμεως μιᾶς. κοινὴ προσκείσθω ἡ λείψις, κὴ ἀπὸ ὁμοίων ὅμοια. δυνάμεις ἄρα ι ἴσαι ἀριθμοῖς ιϛ. κὴ γίνεται ὁ ἀριθμὸς ιϛ. πέμπτων. ἴσται ὁ μὲν σϛ εἰκοσιπέμπτων. ὁ δὲ μυδ' εἰκοσιπέμπτων, ὧ οἱ δύο συντιθέντες ποιοῦσι

ῡ εἰκοσιπέμπτα, ἤτοι μονάδας ιϛ. καὶ ἔστιν ἑκάτερος τετράγωνος.

OBSERVATIO DOMINI PETRI DE FERMAT.

CVbum autem in duos cubes, aut quadratoquadratum in duos quadratoquadratos & generaliter nullam in infinitum vltra quadratum poteftatem in duos eiufdem nominis fas est diuidere cuius rei demonstrationem mirabilem fane detexi. Hanc marginis exiguitas non caperet.

Fig 29 Fermat's marginal note, published in his son's edition of the *Arithmetica* of Diophantus.

examples, a tradition that goes back to ancient Babylon, and he didn't provide proofs.

Fermat's personal copy of the *Arithmetica* has not survived, but he must have written his marginal note in it, because Samuel says so. Fermat is unlikely to have left such a treasure unopened for very long, and his conjecture is such a natural one that he probably thought of it as soon as he read question VIII of Book II. He evidently wondered whether anything similar could be achieved using cubes instead of squares, a natural question for a mathematician to ask. He found no examples – we can be sure of that since none exist – and he was equally unsuccessful when he tried higher powers, for example fourth powers. He decided that these questions had no solutions. His marginal note says as much; it translates into English as:

It is impossible to divide a cube into two cubes, or a fourth power into two fourth powers, or in general, any power higher than the second, into two like powers. I have discovered a truly marvellous proof of this, which this margin is too narrow to contain.

In algebraic language, Fermat was claiming to have proved that the Diophantine equation

$$x^n + y^n = z^n$$

has no whole number solutions if n is any integer greater than or equal to 3. Clearly he was ignoring trivial solutions in which x or y is zero. To avoid repeating the formula all over the place, I'll refer to it as the Fermat equation.

If Fermat really did have a proof, no one has ever found it. The theorem was finally proved true in 1995, more than three and a half centuries after he first stated it, but the methods go way beyond anything that was available in his day, or that he could have invented. The search for a proof had a huge influence on the development of mathematics. It pretty much caused the creation of algebraic number theory, which flourished in the nineteenth century because of a failed attempt to prove the theorem and a brilliant idea that partially rescued it. In the late twentieth and twenty-first centuries, it sparked a revolution.

Early workers on Fermat's last theorem tried to pick off powers one by one. Fermat's general proof, alluded to in his margin, may or may not have existed, but we do know how he proved the theorem for fourth powers. The main tool is Euclid's recipe for Pythagorean triples. The fourth power of any number is the square of the square of that number. So any solution of the Fermat equation for fourth powers is a Pythagorean triple, for which all three numbers are themselves squares. This extra condition can be plugged into Euclid's recipe, and after some cunning manoeuvres, what emerges is *another* solution to the Fermat equation for fourth powers.[52] This might not seem like progress; after a page of algebra, the problem is reduced to the same problem. However, it really is reduced: the numbers in the second solution are smaller than those in the first, hypothetical, solution. Crucially, if the first solution is not trivial – if x and y are nonzero – then the same is true of the second solution. Fermat pointed out that repeating this procedure would lead to a sequence of solutions in which the numbers become perpetually smaller. However, any decreasing sequence of whole numbers must eventually stop. This is a logical contradiction, so the hypothetical solution does not exist. He called this method 'infinite descent'. We now recognise it as a proof by mathematical induction, mentioned in chapter 4, and it can be rephrased in terms of minimal criminals. Or, in this case, minimal models of virtue. Suppose there exists a virtuous citizen, a nontrivial solution of the equation. Then there exists a minimal virtuous citizen. But Fermat's argument then implies the existence of an even smaller virtuous citizen – contradiction. Therefore no citizens can be virtuous. Different proofs for fourth powers have been appearing ever since, and around 30 are now known.

Fermat exploited the simple fact that a fourth power is a special kind of square. The same idea shows that in order to prove Fermat's last theorem, it can be assumed that the power n is either 4 or an odd prime. Any number n greater than two is divisible by 4 or by an odd prime p, so every nth power is either a fourth power or a pth power. Over the next two centuries, Fermat's last theorem was proved for exactly three odd primes: 3, 5, and 7. Euler dealt with cubes in 1770; although there is a gap in the published proof, it can be filled using a result that Euler published elsewhere. Legendre and Peter Lejeune-Dirichlet dealt with fifth powers around 1825. Gabriel Lamé proved Fermat's last theorem for seventh powers in 1839. Many different proofs were later found for these cases. Along the way, several mathematicians developed proofs when the power is 6, 10, and 14, but these were superseded by the proofs for 3, 5, and 7.

Each proof makes extensive use of algebraic features that are special to the power concerned. There was no hint of any general structure that might prove the theorem for all powers, or even for a significant number of different powers. As the powers got bigger, the proofs became ever more complicated. Fresh ideas were needed, and they had to break new ground. Sophie Germain, one of the great woman mathematicians, divided Fermat's last theorem for a prime power p into two subcases. In the first case, none of the numbers x, y, z is divisible by p. In the second case, one of them is. By considering special 'auxiliary' primes related to p she proved that the first case of Fermat's last theorem has no solutions for an odd prime power less than 100. However, it was difficult to prove much about auxiliary primes in general.

Germain corresponded with Gauss, at first using a masculine pseudonym, and he was very impressed by her originality. When she revealed she was a woman, he was even more impressed, and said so. Unlike many of his contemporaries, Gauss did not assume that women were incapable of high intellectual achievement, in particular mathematical research. Later, Germain made an unsuccessful attempt to prove the first case of Fermat's last theorem for all even powers, where again it is possible to exploit Euclid's characterisation of Pythagorean triples. Guy Terjanian finally disposed of the even powers in 1977. The second case seemed a much harder nut to crack, and no one got very far with it.

In 1847 Lamé, moving on from his proof for seventh powers, had a wonderful idea. It required the introduction of complex numbers, but by that time everyone was happy with those. The vital ingredient was the same one that Gauss had exploited to construct a regular 17-sided polygon, chapter 3. Every number theorist knew about it, but until Lamé, no one had seriously wondered whether it might be just the job for proving Fermat's last theorem.

In the system of real numbers, 1 has exactly one pth root (when p is odd), namely 1 itself. But in complex numbers, 1 has many pth roots; in fact, exactly p of them. This fact is a consequence of the fundamental theorem of algebra, because these roots satisfy the equation $x^p - 1 = 0$, which has degree p. There is a nice formula for these complex pth roots of unity, as they are called, and it shows that they are the powers $1, \zeta, \zeta^2, \zeta^3, \ldots, \zeta^{p-1}$ of a particular complex number ζ.[53] The defining property of these numbers implies that $x^p + y^p$ splits into p factors:

$$x^p + y^p = (x + y)(x + \zeta y)\left(x + \zeta^2 y\right) \cdots \left(x + \zeta^{p-1} y\right)$$

By the Fermat equation, this expression is also equal to z^p, which is the pth power of an integer. Now, it is easy to see that if a product of numbers, having no common factor, is a pth power, then each number is itself a pth power. So, give or take a few technicalities, Lamé could write each factor as a pth power. From this he deduced a contradiction.

Lamé announced the resulting proof of Fermat's last theorem to the Paris Academy in March 1847, giving credit for the basic idea to Joseph Liouville. Liouville thanked Lamé, but pointed out a potential issue. The crucial statement implying that each factor is a pth power is not a done deal. It depends on uniqueness of prime factorisation – not just for ordinary integers, where that property is true, but for the new kinds of number that Lamé had introduced. These combinations of powers of ζ are called cyclotomic integers; the word means 'circle-cutting', and refers to the connection that Gauss had exploited. Not only was the property of unique prime factorisation not proved for cyclotomic integers, said Liouville: it might be false.

Others already had doubts. Three years earlier, in a letter, Gotthold Eisenstein wrote:

> If one had the theorem which states that the product of two complex numbers can be divisible by a prime number only when one of the factors is – which seems completely obvious – then one would have the whole theory [of algebraic numbers] at a single blow; but this theorem is totally false.

The theorem to which he alludes is the main step needed for a proof of uniqueness of prime factorisation. Eisenstein was referring not just to the numbers that Lamé needed, but to similar numbers arising from other equations. They are called algebraic numbers. An algebraic number is a complex number that satisfies a polynomial equation with rational coefficients. An algebraic integer is a complex number that satisfies a polynomial equation with integer coefficients, provided the coefficient of the highest power of x is 1. For each such polynomial, we obtain an associated algebraic number field (meaning that you can add, subtract, multiply, and divide such numbers to get numbers of the same kind) and its ring (similar but omit 'divide') of algebraic integers. These are the basic objects that are studied in algebraic number theory.

If, for instance, the polynomial is $x^2 - 2$, then it has a solution $\sqrt{2}$. The field consists of all numbers $a + b\sqrt{2}$ with a, b rational; the ring of integers consists of the numbers of this form with a, b integers. Again prime factors can be defined, and are unique. There are some surprises: the polynomial $x^2 + x - 1$ has a solution $(\sqrt{5} - 1)/2$, so despite the fraction, this is an algebraic *integer*.

In algebraic number theory, the difficulty is not to define factors. For example, a cyclotomic integer is a factor of (that is, divides) another if the second is equal to the first multiplied by some cyclotomic integer. The difficulty is not to define primes: a cyclotomic integer is prime if it has no factors, other than trivial 'units', which are the cyclotomic integers that divide 1. There is no problem about resolving a cyclotomic integer, or any other algebraic number, into prime factors. Just keep factorising it until you run out of factors. There is a simple way to prove that the procedure stops, and when it does, every factor must be prime. So what's the difficulty? Uniqueness. If you run the procedure again, making different choices along the way, it might stop with a different list of prime factors.

At first sight, it's hard to see how this can happen. The prime factors are the smallest possible pieces into which the number can be split. It's like taking a Lego toy and pulling it apart into its component bricks. If there were another way to do that, it would end up pulling one of those bricks apart into two or more pieces. But then it wouldn't be a brick. Unfortunately, the analogy with Lego is misleading. Algebraic numbers aren't like that. They are more like bricks with movable links, able to lock together in different ways. Break up a brick in one way, and the resulting pieces lock together and can't be pulled apart any further. Break it up in a different way, and again the resulting pieces lock together. But now they are different.

I'll give you two examples. The first uses only ordinary integers; it's easy to grasp but it has some unrepresentative features. Then I'll show you a genuine example.

Suppose we lived in a universe where the only numbers that existed were 1, 5, 9, 13, 17, 21, 25, and so on – numbers which in our actual universe have the form $4k + 1$. If you multiply two such numbers together you get another number of the same kind. Define such a number to be 'prime' if it is not the product of two smaller numbers *of that kind*. For example, 25 is not prime because it is 5×5, and 5 is a number in the list. But 21 *is* prime, in this new sense, because its usual factors 3 and 7 are not in the list. They are of the form $4k + 3$, not $4k + 1$. It is easy to see that every number of the specified kind is a product of primes in the new sense. The reason is that factors, if they exist, must get smaller. Eventually, the process of factorisation has to stop. When it does, the factors concerned are primes.

However, this type of prime factorisation is not unique. Consider the number 4389, which is $4 \times 1097 + 1$, so is of the required form. Here are three distinct factorisations into numbers of the required form:

$$4389 = 21 \times 209 = 33 \times 133 = 57 \times 77$$

I claim that, with our current definition, all of these factors are primes. For example, 57 is prime, because its usual factors 3 and 19 are not of the required form. The same goes for 21, 33, 77, 133, and 209. Now we can explain the lack of uniqueness. In ordinary integers

$$4389 = 3 \times 7 \times 11 \times 19$$

and all of these factors have the 'wrong' form, $4k + 3$. The three different prime factorisations, in the new sense, arise by grouping these numbers in pairs:

$$(3 \times 7) \times (11 \times 19) \quad (3 \times 11) \times (7 \times 19) \quad (3 \times 19) \times (7 \times 11)$$

We need to use pairs because two numbers of the form $4k + 3$, multiplied together, yield a number of the form $4k + 1$.

This example shows that the argument 'the factors must be unique because they are the smallest pieces' doesn't work. It's true that there are *smaller* pieces hanging around ($21 = 3 \times 7$, for instance) but those pieces aren't in the system concerned. The main reason why this example isn't entirely representative is that although multiplying together numbers of the form $4k + 1$ produces numbers of the same form, that's not true for addition. For instance, $5 + 5 = 10$ is not of the required form. So, in the jargon of abstract algebra, we're not working in a ring.

The second example doesn't have that defect, but in compensation, it's a bit harder to analyse. It is the ring of algebraic integers for the polynomial $x^2 - 15$. This ring consists of all numbers $a + b\sqrt{15}$ where a and b are integers. In it, the number 10 has two distinct factorisations:

$$10 = 2 \times 5 = (5 + \sqrt{15}) \times (5 - \sqrt{15})$$

All four factors 2, 5, $5 + \sqrt{15}$, $5 - \sqrt{15}$ can be proved prime.[54]

All this is much clearer now than it was in 1847, but it didn't take long to show that Liouville's doubts were justified. A fortnight after he expressed them, Wantzel informed the academy that uniqueness was true for some small values of p, but his method of proof failed for 23rd powers. Shortly afterwards, Liouville told the academy that unique prime factorisation is *false* for cyclotomic integers corresponding to $p = 23$. Ernst Kummer had discovered this three years earlier, but hadn't told anyone because he was working on a method for getting round the obstacle. Lamé's proof worked for smaller values of p, including some new ones: 11, 13, 17, 19. But for the general case, the proof was in tatters. It was an object lesson in not assuming that plausible mathematical statements are obvious. They may not even be true.

Kummer had been thinking about Fermat's last theorem, along similar lines to Lamé's. He noticed the potential obstacle, took it seriously, investigated it, and discovered that it wrecked that approach. He found an explicit example of non-unique prime factorisation for cyclotomic integers based on 23rd roots of unity. But Kummer was not one to give up easily, and he found a way to get round the obstacle – or, at least, to mitigate its worst effects. His idea is especially transparent in the case of those $4k + 1$ numbers. The way to restore unique factorisation is to throw in some *new* numbers, outside the system we're interested in. For that example, what we need are the missing $4k + 3$ numbers. Or we can go the whole hog and throw in the even integers as well; then we get the integers, which are closed under addition and multiplication. That is, if you add or multiply two integers, the result is an integer.

Kummer came up with a version of the same idea. For instance, we can restore unique prime factorisation in the ring of all numbers $a + b\sqrt{15}$ by throwing in a new number, namely $\sqrt{5}$. To obtain a ring, it turns out that we must also throw in $\sqrt{3}$. Now

$$2 = (\sqrt{5} + \sqrt{3}) \times (\sqrt{5} - \sqrt{3}) \qquad 5 = \sqrt{5} \times \sqrt{5}$$

and

$$5 + \sqrt{15} = \sqrt{5} \times (\sqrt{5} + \sqrt{3}) \qquad 5 - \sqrt{15} = \sqrt{5} \times (\sqrt{5} - \sqrt{3})$$

So the two factorisations arise by grouping the four numbers $\sqrt{5}$, $\sqrt{5}$, $\sqrt{5} + \sqrt{3}$, $\sqrt{5} - \sqrt{3}$ in two different ways.

Kummer called these new factors ideal numbers, because in his general formulation they weren't exactly numbers at all. They were symbols that behaved a lot like numbers. He proved that every cyclotomic integer can be factorised uniquely into prime ideal numbers. The set-up was subtle: neither the cyclotomic integers, nor the ideal numbers, had unique prime factorisation. But if you used the ideal numbers as the ingredients for prime factorisation of cyclotomic integers, the result was unique.

Later Richard Dedekind found a more civilised reinterpretation of Kummer's procedure, and this is the one we now use. To each ideal number outside the ring concerned, he associated a *set* of numbers inside the ring. He called such a set an ideal. Every number in the ring defines an ideal: it consists of all multiples of that number. If prime factorisation is unique, every ideal is like that. When it's not, there are extra ideals. We can define product and sums of ideals, and prime ideals, and Dedekind proved that prime factorisation *of ideals* is unique for all rings of algebraic integers. This suggests that for most problems you should work with ideals, not the algebraic numbers themselves. Of course, that introduces new complexities, but the alternative is usually to get stuck.

Kummer was able to work with his ideal numbers – well enough to prove a version of Fermat's last theorem with some extra hypotheses. But other mortals found ideal numbers rather difficult, if not a bit mystical. Once viewed in Dedekind's way, however, ideal numbers made perfect sense, and algebraic number theory took off. One important idea that emerged was a way to measure how badly unique factorisation fails in a ring of algebraic integers. To each such ring there corresponds a whole number called its *class number*. If the class number is 1, prime factorisation is unique; otherwise it's not. The bigger the class number, the 'less unique' prime factorisation is, in a meaningful sense.

Being able to quantify the lack of uniqueness was a big step forward, and with extra effort it rescued Lamé's strategy – sometimes. In 1850 Kummer announced that he could prove Fermat's last theorem for a great many primes, those that he called regular. Among the primes up to 100, only 37, 59, and 67 are irregular. For all other primes up to that limit, and many beyond it, his methods proved Fermat's last theorem. The definition of a regular prime requires the class number: a prime is regular if it does not divide the class number of the corresponding ring of cyclotomic integers. So for a regular prime, although prime factorisation is not unique, the way it fails to be unique does not involve the prime concerned in an essential manner.

Kummer claimed that there exist infinitely many regular primes, but this assertion remains unproved. Ironically, in 1915 K.L. Jensen proved that there are infinitely many irregular primes. A bizarre criterion for a prime to be regular emerged from connections with analysis. It involves a sequence of numbers discovered independently by the Japanese mathematician Seki Takazu (or Kōwa) and the Swiss mathematician Jacob Bernoulli, called Bernoulli numbers. This criterion shows that the first ten irregular primes are 37, 59, 67, 101, 103, 131, 149, 157, 233, and 257. By digging more deeply into the structure of cyclotomic integers, Dmitri Mirimanoff disposed of the first irregular prime, 37, in 1893. By 1905 he had proved Fermat's last theorem up to $p = 257$. Harry Vandiver developed computer algorithms that extended this limit. Using these methods, John Selfridge and Bary Pollack proved the theorem up to the 25,000th power in 1967, and S. Wagstaff increased that to 100,000 in 1976.

The evidence for the truth of Fermat's last theorem was piling up, but the main implication was that if the theorem were false, then a counterexample – an example exhibiting its falsity – would be so gigantic that no one would ever be able to find it. The other implication was that methods like Kummer's were running into the same problems that afflicted the work of the earlier pioneers: bigger powers required special, complicated treatment. So this line of attack slowly ground to a halt.

When you get stuck on a mathematical problem, Poincaré's advice is spot on: go away and do something else. With luck and a following wind, a new idea will eventually turn up. Number theorists didn't consciously follow his advice, but nevertheless, they did what he had recommended. As Poincaré had insisted it would, the tactic worked. Some number theorists turned their attention to elliptic curves, chapter 6. Ironically, this area eventually turned out to have a startling and unexpected link to Fermat's last theorem, leading to Wiles's proof. To describe this link, one further concept is required: that of a modular function. The discussion is going to get a bit technical, but there is a sensible story behind the ideas and all we'll need is a broad outline. Bear with me.

In chapter 6 we saw that the theory of elliptic functions had a profound effect on complex analysis. In the 1830s Joseph Liouville discovered that the variety of elliptic functions is fairly limited. Given the two periods, there is a special elliptic function, the Weierstrass function, and every other elliptic function with those two periods is a simple variant. This implies that the only doubly periodic functions you needed to understand are the Weierstrass functions – one for each pair of periods.

Geometrically, the doubly periodic structure of an elliptic function can be phrased in terms of a lattice in the complex plane: all integer combinations $mu + nv$ of the two periods u and v, for integers m and n, Figure 30. If we take a complex number z and add one of these lattice points to it, the elliptic function at this new point has the same value that it had at the original point. In other words, the elliptic function has the same symmetry as the lattice.

Analysts had discovered a much richer source of symmetries of the complex plane, known as Möbius transformations. These change z to $(az + b)/(cz + d)$, for complex constants a, b, c, d. Lattice symmetries are special kinds of Möbius transformation, but there are others. Sets of points analogous to the lattice still exist in this more general setting. A lattice defines a tiling pattern in the Euclidean plane: use a parallelogram for the tiles and place its corners at the lattice points, Figures 26 and 30. Using Möbius transformations, we can construct tiling patterns in a suitable non-Euclidean geometry, the hyperbolic plane. We can identify this geometry with a region of the complex plane, in which straight lines are replaced by arcs of circles.

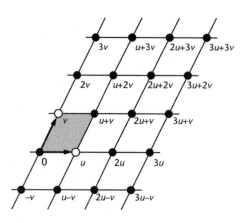

Fig 30 The lattice is formed from all integer combinations of the two periods.

There are highly symmetric tiling patterns in hyperbolic geometry. For each of them, we can construct complex functions that repeat the same values on every tile. These are known as modular functions, and they are natural generalisations of elliptic functions. Hyperbolic geometry is a very rich subject, and the range of tiling patterns is much more extensive than it is for the Euclidean plane. So complex analysts started thinking seriously about non-Euclidean geometry. A profound link between analysis and number theory then appeared. Modular functions do for elliptic curves what trigonometric functions do for the circle.

Recall that the unit circle consists of points (x,y) such that $x^2 + y^2 = 1$. Suppose A is a real number, and set

$$x = \cos A \qquad y = \sin A$$

Then the definition of sine and cosine tells us that this point lies on the unit circle. Moreover, every point on the unit circle is of this form. In the jargon, the trigonometric functions *parametrise* the circle. Something very similar happens for modular functions. If we define x and y using suitable modular functions of a parameter A, the corresponding point lies on an elliptic curve – the same elliptic curve, whatever value A takes. There are more abstract ways to make this statement precise, and workers in the area use those because they are more convenient, but this version brings out the analogy with trigonometry and the circle. This connection produces an elliptic curve for each modular function, and the variety of modular functions is huge – all symmetric tilings of the hyperbolic plane. So an awful lot of elliptic curves can be related to modular functions. Which elliptic curves can be obtained in this way? That turned out to be the heart of the matter.

This 'missing link' first came to prominence in 1975, when Yves Hellegouarch noticed a curious connection between Fermat's last theorem and elliptic curves. Gerhard Frey developed the idea further in two papers published in 1982 and 1986. Suppose, as always, that p is an odd prime. Assume – hoping to derive a contradiction – that there exist nonzero integers a, b, and c that satisfy the Fermat equation, so $a^p + b^p = c^p$. Now extract the rabbit from the hat with a theatrical flourish: consider the elliptic curve

$$y^2 = x(x - a^p)(x - b^p)$$

This is called the Frey elliptic curve. Frey applied the machinery of elliptic curves to this one, and what emerged was a string of ever more bizarre coincidences. His hypothetical elliptic curve is very strange indeed. It doesn't seem to make sense. Frey proved that it makes so little sense that it can't exist. And that, of course, proves Fermat's last theorem, by providing the required contradiction.

However, there was a gap, one that Frey was well aware of. In order to prove that his hypothetical elliptic curve does not exist, you have to show that if it did exist, it would be modular – that is, one of the curves that arise from modular functions. We've just seen that such curves are common, and at that time no one had ever found an elliptic curve that was *not* modular. It seemed likely that the Frey curve should be modular – but it was a hypothetical curve, the numbers a, b, c were not known, and if the curve *were* modular, then it wouldn't exist at all. However, there was one way to deal with all those issues: prove that *every* elliptic curve is modular. Then the Frey curve, hypothetical or not, would have to be modular if it existed. And if it didn't exist, the proof was complete anyway.

The statement that every elliptic curve is modular is called the Taniyama-Shimura conjecture. It is named for two Japanese mathematicians, Yutaka Taniyama and Goro Shimura. They met by accident, both wanting to borrow the same book from the library at the same time for the same reason. This triggered a long collaboration. In 1955 Taniyama was at a mathematics conference in Tokyo, and the younger participants were invited to collect together a list of open questions. Taniyama contributed four, all hinting at a relationship between modular functions and elliptic curves. He had calculated some numbers associated with a particular modular function, and noticed that the exact same numbers turned up in connection with a particular elliptic curve. This kind of coincidence is often a sign that it's not actually a coincidence at all; that there must be some sensible explanation. The equality of these numbers is now known to be equivalent to the elliptic curve being modular; in fact, that's the preferred definition in the research literature. Anyway, Taniyama was sufficiently intrigued that he calculated the numbers for a few other modular functions, finding that these, too, correspond to specific elliptic curves.

He began to wonder whether something similar would work for every elliptic curve. Most workers in the field considered this too good to be true, a pipedream for which there was very little evidence. Shimura was one of the few who felt that the conjecture had merit. But in 1957–8 Shimura visited Princeton for a year, and while he was away, Taniyama committed suicide. He left a note that read, in part, 'As to the cause of my suicide, I don't quite understand it myself, but it is not the result of a particular incident, nor of a specific matter. Merely may I say, I am in the frame of mind that I lost confidence in my future.' At the time he had been planning to get married, and the prospective bride, Misako Suzuki, killed herself about a month later. Her suicide note included 'Now that he is gone, I must go too in order to join him.'

Shimura continued to work on the conjecture, and as evidence in favour accumulated, he started to think it might actually be true. Most other workers in the area disagreed. Simon Singh[55] reports an interview with Shimura, in which he recalled trying to explain this to one of his colleagues:

> The professor enquired, 'I hear that you propose that some elliptic equations can be linked to modular forms.'

'No, you don't understand,' replied Shimura. 'It's not just *some* elliptic equations, it's *every* elliptic equation!'

Despite this kind of scepticism, Shimura persevered, and after many years the proposal became sufficiently respectable to be referred to as the Taniyama-Shimura conjecture. Then André Weil, one of the great number theorists of the twentieth century, found a lot more evidence in the conjecture's favour, publicised it, and expressed the belief that it might well be true. It became known as the Taniyama-Shimura-Weil conjecture. The name never quite settled down, and many permutations of subsets of the three mathematicians have been associated with it. I will stick to 'Taniyama-Shimura conjecture'.

In the 1960s another heavyweight, Robert Langlands, realised that the Taniyama-Shimura conjecture could be viewed as just one element in a much broader and more ambitious programme, which would unify algebraic and analytic number theory. He formulated a whole raft of conjectures related to this idea, now known as the Langlands programme. It was even more speculative than the Taniyama-Shimura conjecture, but it had a compelling elegance, the sort of mathematics that ought to be true because it was so beautiful. Throughout the 1970s the mathematical world became accustomed to the wild beauty of the Langlands programme, and it started to become accepted as one of the core aims of algebraic number theory. The Langlands programme seemed to be the right way forward, if only someone could take the first step.

At this point Frey noticed that applying the Taniyama-Shimura conjecture to his elliptic curve would prove Fermat's last theorem. However, by then another problem with Frey's idea had emerged. When he gave a lecture about it in 1984, the audience spotted a gap in his key argument: the curve is so bizarre that it can't be modular. Jean-Pierre Serre, one of the leading figures in the area, quickly filled the gap, but he had to invoke another result that also lacked a proof, the special level reduction conjecture. By 1986, however, Ken Ribet had proved the special level reduction conjecture. Now the only obstacle to a proof of Fermat's last theorem was the Taniyama-Shimura conjecture, and the consensus began to shift. Serre predicted that Fermat's last theorem would probably be proved within a decade or so. Exactly how was another matter, but there was a general feeling in the air: the techniques related to modular functions were becoming so powerful that someone would soon make Frey's approach work.

That someone was Andrew Wiles. In the television programme about his proof, he said:

> I was a ten-year-old and ... I found a book on math and it told a bit about the history of this problem [Fermat's last theorem] – that someone had [posed] this problem 300 years ago, but no one had ever seen a proof, no one knew if there was a proof, and people ever since have looked for the proof. And here was a problem that I, a ten-year-old, could understand, but none of the great mathematicians of the past had been able to resolve. And from that moment of course I just tried to solve it myself. It was such a challenge, such a beautiful problem.

In 1971 Wiles took a mathematics degree at Oxford, moving to Cambridge for his PhD. His supervisor, John Coates, advised him (correctly) that Fermat's last theorem was too difficult for a PhD. So instead, Wiles set to work on elliptic curves, then considered to be a far more promising area of research. By 1985–6 he was in Paris at the Institut des Hautes Études Scientifiques (Institute of Advanced Scientific Studies), one of the world's leading mathematical research institutes. Most top researchers pass through it at some point; if you are a mathematician, it's a great place to hang out. Among the visitors was Ribet, and his proof of the special level reduction conjecture electrified Wiles. Now he could pursue entirely respectable research into elliptic curves, by trying to prove the Taniyama-Shimura conjecture, and at the same time he could try to fulfil his childhood dream of proving Fermat's last theorem.

Because everyone in the area now knew of the connection, there was a worry. Suppose Wiles managed to put together an almost complete proof, with a few small gaps that needed extra work. Suppose someone else learned of this, and filled the gap. Then technically, this person would be the one who had proved Fermat's last theorem. Mathematicians generally manage not to behave like that, but when the prize is so great, it is wise to take precautions. So Wiles carried out his research in secret, something that mathematicians seldom do. It wasn't that he didn't trust his colleagues. It was just that he couldn't take the tiniest risk of being pipped at the post.

He laboured for seven years, tucked away in the roof of his house where there was an office. Only his wife and his head of department knew what he was working on. In peace and seclusion, he attacked the problem with every technique he could learn, until the castle walls began to shake under the onslaught. In 1991 Coates put him on to some new results proved by Mattheus Flach. The crack in the wall began to widen ever more rapidly as the siege took its toll.

By 1993 the proof was complete. Now it had to be revealed to the world. Still cautious, Wiles didn't want to risk proclaiming his solution only for some mistake to surface – something that happened to Yoichi Miyaoka in 1988, whose claim of a proof hit the media, only for a fatal error to be found. So Wiles decided to deliver a series of three lectures at the Isaac Newton Institute in Cambridge, a newly founded international research centre for mathematics. The title was innocuous and technical: 'Modular forms, elliptic curves, and Galois theory.' Few were fooled: they knew Wiles was on to something big.

In the third lecture, Wiles outlined a proof of a special case of the Taniyama-Shimura conjecture. He had discovered that something a little less ambitious would also work. Prove that the Frey curve, if it exists, must belong to a special class of elliptic curves, the 'semistable' ones, and prove that all curves *in that class* must be modular. Wiles then proved both results. At the end of the lecture, he wrote a corollary – a supplementary theorem that follows directly from whatever has just been proved – on the blackboard. The corollary was Fermat's last theorem.

When Shimura heard about Wiles's announcement, his comment was brief and to the point. 'I told you so.'

If only it had been that straightforward. But fate had a twist in store. The proof still had to be refereed by experts, and as usual that process turned up a few points that needed further explanation. Wiles dealt with most of these comments, but one of them forced a rethink. Late in 1993 he issued a statement saying that he was withdrawing his claim until he could patch up a logical gap that had emerged. But now he was forced to operate in the full glare of publicity, exactly what he had hoped to avoid.

By March 1994, no repaired proof had appeared, and Faltings expressed a widespread view in the mathematical community: 'If [repairing the proof] were easy, he'd have solved it by now. Strictly speaking, it was not a proof when it was announced.' Weil remarked 'I believe he has some good ideas ... but the proof is not there ... proving Fermat's last theorem is like climbing Everest. If a man wants to climb Everest and falls short of it by 100 yards, he has not climbed Everest.' Everyone could guess how it was going to end. They'd seen it all before. The proof had collapsed, it would have to be retracted completely, and Fermat's last theorem would live to fight another day.

Wiles refused to concede defeat, and his former student Richard Taylor joined the quest. The root of the difficulty was now clear: Flach's results weren't quite suited to the task. They tried to modify Flach's methods, but nothing seemed to work. Then, in a flash of inspiration, Wiles suddenly understood what the obstacle was. 'I saw that the thing that had stopped [Flach's method] working was something that would make another method I had tried previously work.' It was as though the soldiers besieging the castle had realised that their battering-ram would never work because the defenders kept dropping rocks on it, but those self-same rocks could be stuffed into a trebuchet and used to break down the door instead.

By April 1995 the new proof was finished, and this time there were no gaps or errors. Publication quickly followed, two papers in the ultra-prestigious *Annals of Mathematics*. Wiles became an international celebrity, was awarded several major prizes and a knighthood ... and went back to his research, carrying on pretty much as before.

The really important feature of Wiles's solution is not Fermat's last theorem at all. As I said, nothing vital hinges on the answer. If someone had found three 100-digit numbers and a 250-digit prime that provided a counterexample to Fermat's claim, then the theorem would have been false, but no crucial area of mathematics would have been in any way diminished. Of course a direct attack by computer would not be able to search through numbers that large, so you would have to be amazingly clever to establish any such thing, but a negative result would not have caused any heartaches.

The real importance of the solution lies in the proof of the semistable case of the Taniyama-Shimura conjecture. Within six years Christophe Breuil, Brian Conrad, Fred Diamond, and Taylor extended Wiles's methods to handle not just the semistable case, but all elliptic curves. They proved the full Taniyama-Shimura conjecture, and number theory would never be the same. From that point on, whenever anyone encountered an elliptic curve, it was guaranteed to be modular, so a host of analytic methods would open up. Already these methods have been used to solve other problems in number theory, and more will turn up in future.

8

Orbital chaos
Three-Body Problem

ACCORDING TO A TIME-HONOURED joke, you can tell how advanced a physical theory is by the number of interacting bodies that it can't handle. Newton's law of gravity runs into problems with three bodies. General relativity has difficulty dealing with two bodies. Quantum theory is over-extended for one body, and quantum field theory runs into trouble with *no* bodies – the vacuum. Like many jokes, this one contains a grain of truth.[56] In particular, the gravitational interaction of a mere three bodies, assumed to obey Newton's inverse square law of gravity, stumped the mathematical world for centuries. It still does, if what you want is a nice formula for the orbits of those bodies. In fact, we now know that three-body dynamics is chaotic – so irregular that it has elements of randomness.

All this is a huge contrast to the stunning success of Newtonian gravitational theory, which explained, among many other things, the orbit of a planet round the Sun. The answer is what Kepler had already deduced empirically from astronomical observations of Mars: an ellipse. Here only two bodies occur: Sun and planet. The obvious next step is to use Newton's law of gravity to write down the equation for three-body orbits, and solve it. But there is no tidy geometric characterisation of three-body orbits, not even a formula in coordinate geometry. Until the late nineteenth century, very little was known about the motion of three celestial bodies, even if one of them were so tiny that its mass could be ignored.

Our understanding of the dynamics of three (or more) bodies has grown dramatically since then. A large part of that progress has been an ever-increasing realisation of how difficult the question is, and why. That may seem a retrograde step, but sometimes the best way to advance is to make a strategic retreat and try something else. For the three-body problem, this plan of campaign has scored some real successes, when a head-on attack would have become hopelessly bogged down.

Early humans cannot have failed to notice that the Moon moves gradually across the night sky, relative to the background of stars. The stars also appear to move, but they do so as a whole, like tiny pinpricks of light on a vast spinning bowl. The Moon is clearly special in another way: it is a great shining disc, which changes shape from new moon to full moon and back again. It's not a pinprick of light like a star.

A few of those pinpricks of light also disobey the rules. They go walkabout. They don't change their position relative to the stars as quickly as the Moon, but even so, you don't have to watch the sky for many nights to see that some of them are moving. Five of these wandering stars are visible to the naked eye; the Greeks called them *planetes* – wanderers. They are, of course, the planets, and the five that have been recognised since ancient times are what we now call Mercury, Venus, Mars, Jupiter, and Saturn – all named after Roman gods. With the aid of telescopes we now know of two more: Uranus and Neptune. Plus our own Earth, of course. No longer does Pluto count as a planet, thanks to a controversial decision on terminology made by the International Astronomical Union in 2006.

As ancient philosophers, astronomers, and mathematicians studied the heavens, they realised that the planets do not just wander around at random. They follow convoluted but fairly predictable paths, and they return to much the same position in the night sky at fairly regular intervals of time. We now explain these patterns as periodic motion round a closed orbit, with a small contribution from the Earth's own orbital motion. We also recognise that the periodicity is not exact – but it comes close. Mercury takes nearly 88 days to circle the Sun, while Jupiter takes nearly 12 years. The further from the Sun the planet is, the longer it takes to complete one orbit.

The first quantitatively accurate model of the motion of the planets was the Ptolemaic system, named for Claudius Ptolemy, who described it in his *Almagest* ('The greatest [treatise]') of about AD 150. It is a geocentric – Earth-centred – model, in which all celestial bodies orbit the Earth. They move as if supported by a series of gigantic spheres, each rotating at a fixed rate about an axis that may itself be supported by another sphere. Combinations of many rotating spheres were required to represent the complex motion of the planets in terms of the cosmic ideal of uniform rotation in a circle – the equator of the sphere. With enough spheres and the right choices for their axes and speeds, the model corresponds very closely to reality.

Nicolaus Copernicus modified Ptolemy's scheme in several ways. The most radical was to make all bodies, other than the Moon, revolve round the Sun instead, which simplified the description considerably. It was a heliocentric model. This proposal fared ill with the Catholic church, but eventually the scientific view prevailed, and educated people accepted that the Earth revolves round the Sun. In 1596 Kepler defended the Copernican system in his *Mysterium Cosmographicum* ('The cosmographic mystery'), whose high point was his discovery of a mathematical relationship between a planet's distance from the Sun and its orbital period. Moving outwards from the Sun, the ratio of the increase in period from one planet to the next is twice the increase in distance. Later, he decided that this relationship was too inaccurate to be correct, but it sowed the seeds of a more accurate relationship in his future work. Kepler also explained the spacing of the planets in terms of the five regular solids, nested neatly inside each other, separated by the spheres that held them. Five solids explained why there were five planets, but we now recognise eight, so this feature is no longer an advantage. There are 120 different ways to arrange five solids in order, and one of these is likely to come close to the celestial proportions given by planetary orbits. So it's just an accidental approximation, shoehorning nature into a meaningless pattern.

In 1600 the astronomer Tycho Brahe hired Kepler to help him analyse his observations, but political problems intervened. After Brahe's death, Kepler was appointed imperial mathematician to Rudolph II. In his spare time, he worked on Brahe's observations of Mars. One outcome was *Astronomia Nova* ('A new astronomy') of 1609, which presented two more laws of planetary motion. Kepler's first law states that planets move in ellipses – he had established this for Mars, and it seemed likely that the same would be true of the other planets. Initially he assumed that an egg shape would fit the data, but that didn't work out, so he tried an ellipse. This, too, was rejected, and he found a different mathematical description of the orbit's shape. Finally he realised that this was actually just another way to define an ellipse:[57]

> I laid [the new definition] aside, and fell back on ellipses, believing that this was quite a different hypothesis, whereas the two, as I shall prove in the next chapter, are one in the same... Ah, what a foolish bird I have been!

Kepler's second law states that the planet sweeps out equal areas in equal times. In 1619, in his *Harmonices Mundi* ('Harmonies of the world'), Kepler completed his three laws with a far more accurate relationship between distances and periods: the cube of the distance (half the length of the major axis of the ellipse) is proportional to the square of the period.

The stage was now set for Isaac Newton. In his *Philosophiae Naturalis Principia Mathematica* ('Mathematical principles of natural philosophy') of 1687, Newton proved that Kepler's three laws are equivalent to a single law of gravitation: two bodies attract each other with a force that is proportional to their masses and inversely proportional to the square of the distance between them. Newton's law had a huge advantage: it applied to any system of bodies, however many there might be. The price to be paid was the way the law prescribed the orbits: not as geometric shapes, but as solutions of a differential equation, which involved the planets' accelerations. It is not at all clear how to find the shapes of planetary orbits, or the planets' positions at a given time, from this equation. It is not terribly clear how to find their accelerations, to be blunt. Nevertheless, the equation *implicitly* provided that information. The problem was to make it explicit. Kepler had already done that for two bodies, and the answer was elliptical orbits pursued with speeds that sweep out areas at a constant rate.

What about three bodies?

It was a good question. According to Newton's law, all bodies in the solar system influence each other gravitationally. In fact, all bodies in the entire universe influence each other gravitationally. But no one in their right mind would try to write down differential equations for every body in the universe. As always, the way forward was to simplify the problem – but not too much. The stars are so far away that their gravitational effect on the solar system is negligible, unless you want to describe how the Sun moves as the galaxy rotates. The motion of the Moon is mainly influenced by two other bodies: the Earth and the Sun, except for some subtle effects involving other planets. In the early 1700s, this question escaped the realms of astronomy and acquired practical significance, when it was realised that the motion of the Moon could be useful for navigation. (No GPS in those days; not even chronometers to measure longitude.) But this method required more accurate predictions than existing theories could provide. The obvious place to start was to write down the implications of Newton's law for three bodies, which for this purpose could be treated as point masses, because the planets are exceedingly small compared to the distances between them. Then you solved the resulting differential equations. However, the tricks that led from two bodies to ellipses failed when an extra body entered the mix. A few preliminary steps worked, but then the calculation hit an obstruction. In 1747 Jean d'Alembert and Alexis Clairaut, bitter rivals, both competed for a prize from the Paris Academy of Sciences on the 'problème des trois corps', which they both approached through numerical approximations. The three-body problem had acquired its name, and it soon became one of the great enigmas of mathematics.

Some special cases could be solved. In 1767 Euler discovered solutions in which all three bodies lie on a rotating straight line. In 1772 Lagrange found similar solutions where the bodies form a rotating equilateral triangle, which expands and contracts. Both solutions were periodic: the bodies repeated the same sequence of movements indefinitely. However, even drastic simplifications failed to produce anything more general. You could assume that one of the bodies had negligible mass, you could assume that the other two moved in perfect circles about their mutual centre of mass, a version known as the 'restricted' three-body problem ... and *still* you couldn't solve the equations exactly.

In 1860 and 1867 the astronomer and mathematician Charles-Eugène Delaunay attacked the specific case of the Sun-Earth-Moon system using perturbation theory, which views the effect of the Sun's gravity on the Moon as a small change imposed on the effect of the Earth, and derived approximate formulas in the form of series: many successive terms added together. He published his results in 1860 and 1867; each volume was 900 pages long, and consisted largely of the formulas. In the late 1970s his calculations were checked using computer algebra, and only two small and unimportant errors were found.

It was a heroic calculation, but the series approached its limiting value too slowly to have much practical use. However, it spurred others to seek series solutions that converged more rapidly. It also uncovered a big technical obstacle to all such approaches, known as small denominators. Some terms in the series are fractions, and the denominator (the part on the bottom) becomes very small if the bodies are near a resonance: a periodic state in which their periods are rational multiples of each other. For example, Jupiter's three innermost moons Io, Europa, and Ganymede have periods of revolution around the planet of 1.77 days, 3.55 days, and 7.15 days, in almost exact 1 : 2 : 4 ratios. Secular resonances, rational relations between the rates at which the axes of two nearly elliptical orbits turn, are an especial nuisance, because the probable error in evaluating a fraction gets very large when its denominator is small.

If the three-body problem was difficult, the *n*-body problem – any number of point masses moving under Newtonian gravity – was surely harder. Yet nature presents us with an important example: the entire solar system. This contains eight planets, several dwarf planets such as Pluto, and thousands of asteroids, many quite large. Not to mention satellites, some of which – Titan, for instance – are larger than the planet Mercury. So the solar system is a 10-body problem, or a 20-body problem, or a 1000-body problem, depending on how much detail you want to include.

For short-term predictions, numerical approximations are effective, and in astronomy, a thousand years is short. Understanding how the solar system will evolve over hundreds of millions of years is quite another matter. And one big question depends on that kind of long-term view: the stability of the solar system. The planets seem to be moving in relatively stable, almost-elliptical orbits. The orbits change a little bit when other planets perturb them, so the period might change by a fraction of a second, or the size of the ellipse might not be exactly constant. Can we be sure that this gentle jostling is all that will happen in future? Is it typical of what happened in the past, especially in the early stages of the solar system? Will the solar system remain stable, or will two planets collide? Could a planet be flung out into the distant reaches of the universe?

The year 1889 was the 60th birthday of Oscar II, king of Norway and Sweden. As part of the celebrations, the Norwegian mathematician Gösta Mittag-Leffler persuaded the king to announce a prize for the solution of the n-body problem. This was to be achieved not by an exact formula – by then it was clear that this was asking too much – but by some kind of convergent series. Poincaré became interested, and he decided to start with a very simple version: the restricted three-body problem, where one body has negligible mass, like a tiny particle of dust. If you apply Newton's law naively to such a particle, the force exerted on it is the product of the masses divided by the square of the distance, and one of the masses is zero, so the product is zero. That's not very helpful, because the dust particle simply goes its own way, decoupled from the other two bodies. Instead, you set up the model so that the dust particle feels the effect of the other two bodies, but they ignore it completely. So the orbits of the two massive bodies are circular, and they move at a fixed speed. All of the complexity of the motion is invested in the dust particle.

Poincaré didn't solve the problem that King Oscar posed. That was just too ambitious. But his methods were so innovative, and made so much progress, that he was awarded the prize anyway. His prizewinning research was published in 1890, and it suggested that even the restricted three-body problem might not possess the kind of answer that had been stipulated. Poincaré divided his analysis into several distinct cases, depending on general features of the motion. In most of them, series solutions might well be obtainable. But there was one case in which the dust particle's orbit became extraordinarily messy.

Poincaré deduced this inescapable messiness from some other ideas he had been developing, which made it possible to describe solutions to differential equations without actually solving them. This 'qualitative theory of differential equations' was the seed from which modern nonlinear dynamics has grown. The basic idea was to explore the geometry of the solutions; more precisely their topology, a topic in which Poincaré was also deeply interested, chapter 10. In this interpretation, the positions and speeds of the bodies are coordinates in a multidimensional space. As time passes, any initial state follows a curved path through this space. The topology of this path, or the entire system of all possible paths, tells us many useful things about the solutions.

A periodic solution, for example, is a path that closes up on itself to form a loop. As time passes, the state goes round and round the loop, repeating the same behaviour indefinitely. The system is then periodic. Poincaré suggested that a good way to detect such loops is to place a multidimensional surface so that it cuts through the loop. We now call it a Poincaré section. Solutions that start on this surface may eventually return to the surface; the loop itself returns to exactly the same point, and solutions through nearby points always return to the section after roughly one period. So a periodic solution can be interpreted as a fixed point of the 'first return map', which tells us what happens to points on the surface when they first come back to it, if they do. This may not seem much of an advance, but it reduces the dimension of the space – the number of variables in the problem. This is almost always a good thing.

Poincaré's great idea starts to come into its own when we pass to the next most complex kind of solution, combinations of several periodic motions. As a simple example, the Earth goes round the Sun roughly every 365 days and the Moon goes round the Earth roughly every 28 days. So the Moon's motion combines these two different periods. Of course the whole point of the three-body problem is that this description is not entirely accurate, but 'quasiperiodic' solutions of this kind are quite common in many-body problems. The Poincaré section detects quasiperiodic solutions: when they return to the surface they don't hit exactly the same point, but the point where they hit moves round and round a closed curve *on the surface*, in small steps.

Poincaré realised that if every solution were like this, he would be able to set up suitable series to model them quantitatively. But when he analysed the topology of the first return map, he noticed that it could be more complicated. Two particular curves, related by the dynamics, might cross each other. That wasn't too bad in itself, but when you pushed the curves along until they hit the surface again, the resulting curves still had to cross – but at a different place. Push them round again, and they crossed yet again. Not only that: these new curves that arose by pushing the original ones along were not actually new. They were parts of the original curves. Sorting out the topology took some clear-headed thinking, because no one had really played this kind of game before. What emerges is a very complex picture like a crazy net, in which the curves repeatedly zigzag to and fro, crossing each other, and the zigzags themselves zigzag to and fro, and so on to any level of complexity. Poincaré in effect declared himself to be stumped:

> When one tries to depict the figure formed by these two curves and their infinity of intersections, each of which corresponds to a doubly asymptotic solution, these intersections form a kind of net, web or infinitely tight mesh ... One is struck by the complexity of this figure that I am not even attempting to draw.

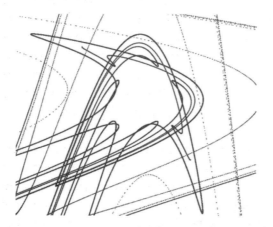

Fig 31 Part of a homoclinic tangle. A complete picture would be infinitely complicated.

We now call his picture a homoclinic ('self-connected') tangle, Figure 31. Thanks to new topological ideas introduced in the 1960s by Stephen Smale, we now recognise this structure as an old friend. Its most important implication is that the dynamics is *chaotic*. Although the equations have no explicit element of randomness, their solutions are very complicated and irregular, sharing certain features of genuinely random processes. For example, there are orbits – most of them, in fact – for which the motion exactly mimics the repeated random tossing of a coin. The discovery that a deterministic system – one whose entire future is uniquely determined by its present state – can nevertheless have random features is remarkable, and it has changed many areas of science. No longer do we automatically assume that simple rules cause simple behaviour. This is what is colloquially known as chaos theory, and it all goes back to Poincaré and his Oscar award.

Well, almost. For many years, historians of mathematics told the story that way. But around 1990 June Barrow-Green found a printed copy of Poincaré's memoir in the depths of the Mittag-Leffler Institute in Stockholm, thumbed through it, and realised that it was different from the version that could be found in innumerable mathematics libraries around the globe. It was, in fact, the official printing of Poincaré's prizewinning memoir, and there was a mistake in it. When Poincaré submitted his work for the prize, he had overlooked the chaotic solutions. He spotted the error before the memoir was published, worked out what he should have deduced – namely chaos – and paid (more than the prize was worth) to have the original version destroyed and a corrected version to be printed. For some reason the Mittag-Leffler Institute archives retained a copy of the original faulty version, but this became forgotten until Barrow-Green unearthed it, publishing her discovery in 1994.

Poincaré seems to have thought that these chaotic solutions were incompatible with series expansions, but that turns out to be wrong too. It was an easy assumption to make: series seem too regular to represent chaos; only topology can do that. Chaos is complicated behaviour caused by simple rules, so the inference isn't watertight, but the structure of the three-body problem definitely precludes simple solutions of the kind that Newton derived for two bodies. The two-body problem is 'integrable', which means that the equations have enough conserved quantities, such as energy, momentum, and angular momentum, to determine the orbits. 'Conserved' means that these quantities do not change as the bodies pursue their orbits. The three-body problem is known not to be integrable.

Even so, series solutions do exist, but they are not universally valid. They fail for initial states with zero angular momentum – a measure of the total spin – which are infinitely rare because zero is a single number among the infinity of all real numbers. Moreover, they are not series in the time variable as such: they are series in its cube root. The Finnish mathematician Karl Fritiof Sundman discovered all this in 1912. Something similar even holds for the n-body problem, again with rare exceptions, a result obtained in 1991 by Qiudong Wang. But for four or more bodies, we don't have any classification of the precise circumstances in which the series fail to converge. We know that such a classification must be very complicated, because there exist solutions where all bodies escape to infinity, or oscillate with infinite rapidity, after a finite time, chapter 12. Physically, these solutions are artefacts of the assumption that the bodies are single (massive) points. Mathematically, they tell us where to look for wild behaviour.

Dramatic progress has been made on the n-body problem when all of the bodies have the same mass. This is seldom a realistic assumption in celestial mechanics, but it is sensible for some non-quantum models of elementary particles. The main interest is mathematical. In 1993 Cristopher Moore found a solution to the three-body problem in which all three bodies play a game of follow-my-leader along the same orbit. Even more surprising is the shape of the orbit: a figure-eight, shown in Figure 32. Although the orbit crosses itself, the bodies never collide.

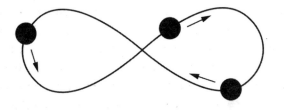

Fig 32 The figure-eight choreography.

Moore's calculation was numerical, on a computer. His solution was rediscovered independently in 2001 by Alain Chenciner and Richard Montgomery, who combined a long-standing principle of classical mechanics, known as 'least action', with some distinctly sophisticated topology to give a rigorous proof that such a solution exists. The orbits are time-periodic: after a fixed interval of time the bodies all return to their initial positions and velocities, and thereafter repeat the same movements indefinitely. For a given common mass, there is at least one such solution for any period.

In 2000 Carles Simó performed a numerical analysis indicating that the figure-eight is stable, except perhaps for a very slow long-term drift known as Arnold diffusion, related to the detailed geometry of Poincaré's return map. For this kind of stability, *almost* all perturbations lead to an orbit very close to the one concerned, and as the perturbation becomes smaller, the proportion of such perturbations approaches 100 per cent. For the small proportion of perturbations that do not behave in this stable manner, the orbit drifts away from its original location extremely slowly. Simó's result was a surprise, because stable orbits are rare in the equal-mass three-body problem. Numerical calculations show that the stability persists, even when the three masses are slightly different. So it is possible that somewhere in the universe, three stars with almost identical masses are chasing each other in a figure-eight. In 2000 Douglas Heggie estimated that the number of such triple stars lies somewhere between one per galaxy and one per universe.

The figure-eight has an interesting symmetry. Start with three bodies A, B, and C. Follow them for one third of the orbital period. You will then find three bodies with the same positions and velocities that they had to begin with – but now the corresponding bodies are B, C, and A. After two thirds of the period the same happens for C, A, and B. A full period restores the original labels of the bodies. This kind of solution is known as a choreography: a planetary dance in which everyone swaps position every so often. Numerical evidence reveals the existence of choreographies for more than three bodies: Figure 33 shows some examples. Simó in particular has found a huge number of choreographies.[58]

Even here, many questions remain unanswered. We lack rigorous proofs of the existence of these choreographies. For more than three bodies they all appear to be unstable; this is most likely correct, but

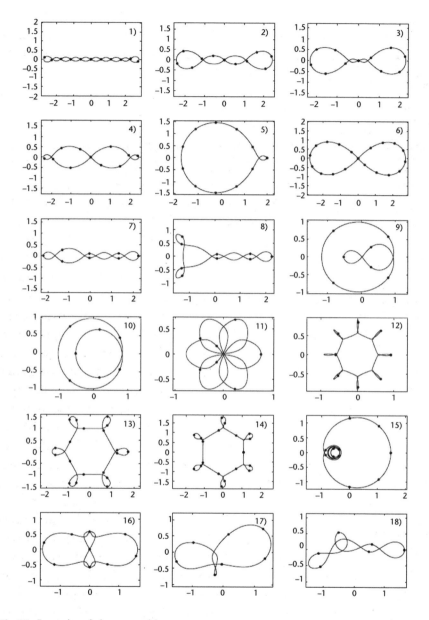

Fig 33 Examples of choreographies.

remains to be proved. The figure-eight orbit for three bodies of given mass and given period appears to be unique, but again no proof is known, although in 2003 Tomasz Kapela and Piotr Zgliczynski provided a computer-aided proof that it is locally unique – no nearby orbit works. Choreographies could be another great problem in the making.

Is the solar system stable, then?

Maybe, maybe not.

By following up Poincaré's great insight, the possibility of chaos, we now understand much more clearly the theoretical issues involved in establishing stability. They turn out to be subtle, complex, and – ironically – not related to the existence of series solutions in any very useful way. Work of Jürgen Moser and Vladimir Arnold has led to proofs that various simplified models of the solar system are stable for almost all initial states, except perhaps for the effect of Arnold diffusion, which prevents stronger kinds of stability in almost all problems of this kind. In 1961 Arnold proved that an idealised model solar system is stable in this sense, but only under the assumption that the planets have very small masses compared to the central star, and the orbits are very close to circular and very close to a common plane. As far as rigorous proof goes, 'very close' here means 'differing by a factor of at most 10^{-43}', and even then, the full statement is that the probability of being unstable is zero. In this kind of perturbation argument, the results are often valid for much larger discrepancies than anything that can be rigorously proved, so the inference is that planetary systems reasonably close to this ideal are probably stable. However, in our solar system the relevant numbers are about 10^{-3} for masses and 10^{-2} for circularity and inclination. Those comfortably exceed 10^{-43}. So the applicability of Arnold's result was moot. It was nevertheless encouraging that *anything* could be said with certainty.

The practical issues in such problems have also become clearer, thanks to the development of powerful numerical methods to solve equations approximately by computer. This is a delicate matter because chaos has an important consequence: small errors can grow very rapidly and ruin the answers. Our theoretical understanding of chaos, and of equations like those for the solar system where there is no friction, has led to the development of numerical methods that are immune to many of the most annoying features of chaos. They are called symplectic integrators. Using them, it turns out that Pluto's orbit is chaotic. However, that doesn't imply that Pluto rushes all over the solar system causing havoc. It means that in 200 million year's time, Pluto will still be somewhere close to its present orbit, but we don't have a clue whereabouts in that orbit it will be.

In 1982 Archie Roy's Project Longstop modelled the outer planets (Jupiter outwards) on a supercomputer, which found no large-scale instability, although some of the planets acquired energy at the expense of others in strange ways. Since then two research groups, in particular, have developed these computational methods and applied them to many different problems about our solar system. They are run by Jack Wisdom and Jacques Laskar. In 1984 Wisdom's group predicted that Saturn's satellite Hyperion, rather than spinning regularly, should tumble chaotically, and subsequent observations confirmed this. In 1988, in collaboration with Gerry Sussman, the group built its own computer, tailored to the equations of celestial mechanics: the digital orrery. An orrery is a mechanical device with cogs and gears, simulating the movement of the planets, which are little metal balls on sticks.[59] The original computation followed the next 845 million years of the solar system, and revealed the chaotic nature of Pluto. With its successors, Wisdom's group has explored the dynamics of the solar system for the next few billion years.

Laskar's group published its first results on the long-term behaviour of the solar system in 1989, using an averaged form of the equations that goes back to Lagrange. Here some of the fine detail is fuzzed out and ignored. The group's calculations showed that the Earth's position in its orbit is chaotic, much like Pluto's: if we measure where the Earth is today, and are wrong by 15 metres, then its position in orbit 100 million years from now cannot be predicted with any certainty.

One way to mitigate the effects of chaos is to perform many simulations, with slightly different initial data, and get a picture of the range of possible futures and how likely each of them is. In 2009 Laskar and Mickaël Gastineau applied this technique to the solar system, following 2500 different scenarios. The differences are extraordinarily small – move Mercury by 1 metre, for example. In about 1 per cent of these futures, Mercury becomes unstable: it collides with Venus, plunges into the Sun, or gets flung out into space.

In 1999 Norman Murray and Matthew Holman investigated the

inconsistency between results like Arnold's, indicating stability, and simulations, indicating instability. 'Are the numerical results incorrect, or are the classical calculations simply inapplicable?' they asked. Using analytic methods, not numerical ones, they demonstrated that the classical calculations don't apply. The perturbations needed to reflect reality are too big. The main source of chaos in the solar system is a near-resonance among Jupiter, Saturn, and Uranus, plus a less important one involving Saturn, Uranus, and Neptune. They also used numerical methods to check this contention, showing that the prediction horizon – a measure of the time it takes small errors to become large enough to have a significant effect – is about 10 million years.[60] Their simulations show that Uranus undergoes occasional near encounters with Saturn, as the eccentricity of its orbit changes chaotically, and there is a chance that it would eventually be ejected from the solar system altogether. However, the likely time is around 10^{18} years. The Sun will blow up into a red giant much sooner, about 5 billion years from now, and this will affect all of the planets, not least because the Sun will lose 30 per cent of its mass. The Earth will move outwards, and might just escape being engulfed by the greatly expanded Sun. However, it is now thought that tidal interactions will eventually pull the Earth into the Sun. The Earth's oceans will have boiled away long before. But since the typical lifetime of a species, in evolutionary terms, is no more than 5 million years, we really don't need to worry about any of these potential catastrophes. Something else will get us first.

The same methods can be used to investigate the solar system's past: use the same equations and just run time backwards, a simple mathematical trick. Until recently astronomers tended to assume the planets have always been close to their present orbits, ever since they condensed out of a cloud of gas and dust surrounding the nascent Sun. In fact, their orbits and composition have been used to infer the size and composition of that primal dust cloud. It now looks as though the planets did not start out in their present orbits. As the dust cloud coalesced under its own gravitational forces, Jupiter – the most massive planet – began to organise the positions of the other bodies, and these in turn influenced each other. This possibility was proposed in 1984 by Julio Fernandez and Wing-Huen Ip, but for a time their work was viewed as a minor curiosity. In 1993 Renu Malhotra started thinking seriously about the way changes in Neptune's orbit might influence the other giant planets, others took up the tale, and a picture of a very dynamic early solar system emerged.

As the planets continued to aggregate, there came a time when Jupiter, Saturn, Uranus, and Neptune were nearly complete, but among them circulated huge numbers of rocky and icy planetesimals, small bodies about 10 kilometres across. From that point on, the solar system evolved through the migration and collision of planetesimals. Many were ejected, which reduced the energy and angular momentum of the four giant planets. Since these worlds had different masses and were at different distances from the Sun, they reacted in different ways. Neptune was one of the winners in the orbital energy stakes, and migrated outwards. So did Uranus and Saturn, to a lesser extent. Jupiter was the big loser, energywise, and moved inwards. But it was so massive that it didn't move very far.

The other, smaller bodies of the solar system were also affected by these changes. Our solar system's current, apparently stable, plan arose through an intricate dance of the giants, in which they threw the smallest bodies at each other in a riot of chaos. So is the solar system stable? Probably not, but we won't be around to find out.

Patterns in primes
Riemann Hypothesis

I N CHAPTER 2 WE LOOKED AT the properties of prime numbers as individuals, and I compared them to the often erratic and unpredictable behaviour of human beings. Humans have free will; they can make their own choices for their own reasons. Primes have to do whatever the logic of arithmetic imposes upon them, but they often seem to have a will of their own as well. Their behaviour is governed by strange coincidences and often lacks any sensible structure.

Nevertheless, the world of primes is not ruled by anarchy. In 1835 Adolphe Quetelet astounded his contemporaries by finding genuine mathematical regularities in social events that depended on conscious human choices or the intervention of fate: births, marriages, deaths, suicides. The patterns were statistical: they referred not to individuals, but to the average behaviour of large numbers of people. This is how statisticians extract order from individual free will. At much the same time, mathematicians began to realise that the same trick works for primes. Although each is a rugged individualist, collectively they conform to the rule of law. There are hidden patterns.

Statistical patterns appear when we think about entire ranges of primes. For instance: how many primes are there up to some specified limit? That's a very difficult question to answer exactly, but there are excellent approximations, and the larger the limit, the better those approximations become. Sometimes the difference between the approximation and the exact answer can be made very small, but usually that's asking too much. Most approximations in this area are asymptotic, meaning that the ratio of the approximation to the exact answer can be made very close to 1. The absolute error in the approximation can grow to any size, even though the percentage error is shrinking towards zero.

If you're wondering how this can be possible, suppose the approximate sequence of numbers, for some abstruse property of primes, is the powers of 100:

$$100 \quad 10,000 \quad 1,000,000 \quad 100,000,000$$

but the exact numbers are

$$101 \quad 10,010 \quad 1,000,100 \quad 100,001,000$$

where the extra 1 moves one place to the left at each stage. Then the ratios of the corresponding numbers get closer and closer to 1, but the differences are

$$1 \quad 10 \quad 100 \quad 1000$$

which become as large as we please. This kind of behaviour happens if the errors – the differences between the approximation and the exact answer – grow without limit, but increase more slowly than the numbers themselves.

The quest for asymptotic formulas related to primes inspired new methods in number theory, based not on whole numbers but on complex analysis. Analysis is the rigorous formulation of calculus, which has two key aspects. One, differential calculus, is about the rate at which some quantity, called a function, changes with respect to another quantity. For example, a body's position depends on – is a function of – time, and the rate at which that position changes as time passes is the body's instantaneous velocity. The other aspect, integral calculus, is about calculating areas, volumes, and the like by adding together large numbers of very small pieces, a process called integration. Remarkably, integration turns out to be the reverse of differentiation. The original formulation of calculus by Newton and Gottfried Leibniz required some manoeuvres with infinitely small quantities, raising questions about the theory's logical validity. Eventually these conceptual issues were sorted out by defining the notion of a limit, a value that can be approximated as closely as required, but need not actually be attained. When presented in this more rigorous way, the subject is called analysis.

In Newton's and Leibniz's day, the quantities concerned were real numbers, and the subject that emerged was real analysis. As complex numbers became widely accepted among mathematicians, it was natural to extend analysis to complex quantities. This subject is complex analysis. It turned out to be extraordinarily beautiful and powerful. When it comes to analysis, complex functions are much better behaved than real ones. They have their peculiarities, mind you, but the advantages of working with complex functions greatly outweigh the disadvantages.

It came as a great surprise when mathematicians discovered that arithmetical features of whole numbers can profitably be reformulated in terms of complex functions. Previously, these two number systems had asked very different questions and used very different methods. But now, complex analysis, an extraordinarily powerful body of techniques, could be used to discover special features of number-theoretic functions; from these, asymptotic formulas and much else could be extracted.

In 1859 a German mathematician, Bernhard Riemann, picked up an old idea of Euler's and developed it in a dramatic new way, defining the so-called zeta function. One of the consequences was an *exact* formula for the number of primes up to some limit. It was an infinite sum, but analysts were accustomed to those. It wasn't just a clever but useless trick; it provided genuine new insights into the primes. There was just one tiny snag. Although Riemann could prove that his formula was exact, its most important potential consequences depended on a simple statement about the zeta function, and Riemann couldn't prove this statement. A century and a half later, we still can't. It's called the Riemann hypothesis, and it is the holy grail of pure mathematics.

In chapter 2 we saw that primes tend to thin out as they get bigger. Since exact results on their distribution seemed out of the question, why not look for statistical patterns instead? In 1797–8 Legendre counted how many primes occur up to various limits, using tables of primes that had recently been provided by Jurij Vega and Anton Felkel.

Vega must have liked lengthy calculations; he constructed tables of logarithms and in 1789 he held the world record for calculating π, to 140 decimal places (126 correct). Felkel just liked calculating primes. His main work is the 1776 *Tafel aller Einfachen Factoren der durch 2, 3, 5 nicht theilbaren Zahlen von 1 bis 10 000 000* ('Table of all prime factors of numbers up to 10 million, except for those divisible by 2, 3, or 5'). There are easy ways to test for factors 2, 3, and 5, mentioned in chapter 2, so he saved a lot of space by omitting those numbers. Legendre discovered an empirical approximation to the number of primes less than a given number x, which is denoted $\pi(x)$. If you've seen π only as a symbol for the number 3.14159, this takes a little getting used to, but it's not hard to work out which is intended, even if you don't notice that they use different fonts. Legendre's 1808 text on number theory stated that $\pi(x)$ seems to be very close to $x/(\log x - 1.08366)$.

In an 1849 letter to the astronomer Johann Encke, Gauss stated that when he was about 15 he wrote a note in his logarithm tables, stating that the number of primes less than or equal to x is $x/\log x$ for large x. As with many of his discoveries, Gauss did not publish this approximation, perhaps because he had no proof. In 1838 Dirichlet informed Gauss of a similar approximation that he had discovered, which boils down to the logarithmic integral function[61]

$$\mathrm{Li}(x) = \int_0^x \frac{dt}{\log t}$$

The ratio of $\mathrm{Li}(x)$ to $x/\log x$ tends to 1 as x becomes large, which implies that if one is asymptotic to $\pi(x)$ then so is the other, but Figure 34 suggests (correctly) that $\mathrm{Li}(x)$ is a better approximation than $x/\log x$. The accuracy of $\mathrm{Li}(x)$ is quite impressive; for example,

$$\pi(1,000,000,000) = 50,847,534$$
$$\mathrm{Li}(1,000,000,000) = 50,849,234.9$$

That of $x/\log x$ is poorer: here it is 48,254,942.4.

The approximation formula – either using $\mathrm{Li}(x)$ or $x/\log x$ – became known as the prime number theorem, where 'theorem' was used in the sense of 'conjecture'. The quest for a proof that these formulas are asymptotic to $\pi(x)$ became one of the key open problems

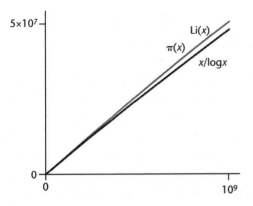

Fig 34 On this scale, $\pi(x)$ and $\text{Li}(x)$ (grey) are indistinguishable. However, $x/\log x$ (black) is visibly smaller. Here x runs horizontally and the value of the function is plotted on the vertical axis.

in number theory. Many mathematicians attacked it using traditional methods of that area, and a few came close; however, there always seemed to be some tricky assumption that evaded proof. New methods were needed. They came from a curious reformulation of two of Euclid's ancient theorems about primes.

The prime number theorem was a response to Euclid's theorem that the primes go on for ever. Another basic Euclidean theorem is the uniqueness of prime factorisation: every positive integer is a product of primes *in exactly one way.* In 1737 Euler realised that the first theorem can be restated as a rather startling formula in real analysis, and the second statement becomes a simple consequence of that formula. I'll start by presenting the formula, and then try to make sense of it. Here it is:

$$\frac{1}{1-2^{-s}} \times \frac{1}{1-3^{-s}} \times \cdots \times \frac{1}{1-p^{-s}} \times \cdots$$

$$= \frac{1}{1^s} + \frac{1}{2^s} + \frac{1}{3^s} + \frac{1}{4^s} + \frac{1}{5^s} + \frac{1}{6^s} + \frac{1}{7^s} + \cdots$$

Here p runs through all the primes and s is constant. Euler was mainly interested in the case when s is a whole number, but his formula works for real numbers as well, provided s is greater than 1. This condition is required to make the series on the right-hand side converge: have a meaningful value when continued indefinitely.

This is an extraordinary formula. On the left-hand side we multiply together infinitely many expressions that depend only on the primes. On the right-hand side we add together infinitely many expressions that depend on all positive whole numbers. The formula expresses, in analytic language, some relation between whole numbers and primes. The main relation of that kind is uniqueness of prime factorisation, and this is what justifies the formula.

I'll sketch the main step to show that there is a sensible idea behind all this. Using school algebra we can expand the expression in p into a series, rather like the right-hand side of the formula but involving only powers of p. Specifically,

$$\frac{1}{1 - p^{-s}} = \frac{1}{1^s} + \frac{1}{p^s} + \frac{1}{p^{2s}} + \frac{1}{p^{3s}} + \cdots$$

When we multiply all of these series together, over all primes p, and 'expand' to obtain a sum of simple terms, we get every combination of prime powers – that is, every whole number. Each occurs as the reciprocal of (1 divided by) its sth power, and each occurs exactly once by uniqueness of prime factorisation. So we get the series on the right.

No one has ever found a simple algebraic formula for this series, although there are many using integrals. So we give it a special symbol, the Greek letter zeta (ζ), and define a new function

$$\zeta(s) = \frac{1}{1^s} + \frac{1}{2^s} + \frac{1}{3^s} + \frac{1}{4^s} + \frac{1}{5^s} + \frac{1}{6^s} + \frac{1}{7^s} + \cdots$$

Euler didn't actually use the symbol ζ, and he considered only positive integer values of s, but I will call the above series the Euler zeta function. Using his formula, Euler deduced that there exist infinitely many primes by allowing s to get very close to 1. If there are finitely many primes, the left-hand side of the formula has a finite value, but the right-hand side becomes infinite. This is a contradiction, so there must be infinitely many primes. Euler's main aim was to obtain formulas like $\zeta(2) = \pi^2/6$, giving the sum of the series for even integers s. He didn't take his revolutionary idea much further.

Other mathematicians spotted what Euler had missed, and considered values of s that are not integers. In two papers of 1848 and 1850 the Russian mathematician Pafnuty Chebyshev had a bright idea: try to prove the prime number theorem using analysis.[62] He started from the link between prime numbers and analysis provided by the Euler zeta function. He didn't quite succeed, because he assumed s to be real, and the analytic techniques available in real analysis were too limited. But he managed to prove that when x is large, the ratio of $\pi(x)$ to $x/\log x$ lies between two constants: one slightly bigger than 1, and one slightly smaller. There was genuine payoff, even with this weaker result, because it allowed him to prove Bertrand's postulate, conjectured in 1845: if you take any integer and double it, there exists a prime between the two.

The stage was now set for Riemann. He also recognised that the zeta function holds the key to the mystery of the prime number theorem, but to make this approach work he had to propose an ambitious extension: define the zeta function not just for a real variable, but for a complex one. Euler's series is a good place to start. It converges for all real s greater than 1, and if exactly the same formula is used for complex s then the series converges whenever the real part of s is greater than 1. However, Riemann discovered he could do much better than that. Using a procedure called analytic continuation he extended the definition of $\zeta(s)$ to *all* complex numbers other than 1. That value of s is excluded because the zeta function becomes infinite when $s = 1$.[63]

In 1859 Riemann put his ideas about the zeta function together into a paper, whose title translates as 'On the number of primes less than a given magnitude'.[64] In it he gave an explicit, exact formula for $\pi(x)$.[65] I'll describe a simpler formula, equivalent to Riemann's, to show how the zeros of the zeta function appear. The idea is to count how many primes or prime powers there are, up to any chosen limit. However, instead of counting each of them once, which is what $\pi(x)$ does for the primes, larger primes are given extra weight. In fact, any power of a prime is counted according to the logarithm of that prime. For example, up to a limit of 12 the prime powers are

$$2, 3, 4 = 2^2, 5, 7, 8 = 2^3, 9 = 3^2, 11$$

so the weighted count is

$$\log 2 + \log 3 + \log 2 + \log 5 + \log 7 + \log 2 + \log 3 + \log 11$$

which is about 10.23.

Using analysis, information about this more sophisticated way to count primes can be turned into information about the usual way. However, this way leads to simpler formulas, a small price to pay for the use of the logarithm. In these terms, Riemann's exact formula states that this weighted count up to a limit x is equal to

$$-\sum_{\rho} \frac{x^{\rho}}{\rho} + x - \tfrac{1}{2}\log(1 - x^{-2}) - \log 2\pi$$

where Σ indicates a sum over all numbers ρ for which $\zeta(\rho)$ is zero, excluding negative even integers. These are called the nontrivial zeros of the zeta function. The trivial zeros are the negative even integers -2, -4, -6, ... The zeta function is zero at these values because of the formula used in the definition of the analytic continuation, but these zeros turn out not to be important for Riemann's formula, or indeed for much else.

In case the formula looks a bit daunting, let me pick out the main point: a fancy way to count primes up to a limit x, which can be turned into the usual way with a bit of analytic trickery, is *exactly* equal to a sum over all nontrivial zeros of the zeta function of the simple expression x^{ρ}/ρ, *plus a straightforward function of x*. If you are a complex analyst, you will see immediately that the prime number theorem is equivalent to proving that the weighted count up to the limit x is asymptotic to x. Using complex analysis, this will be true if all nontrivial zeros of the zeta function have real parts between 0 and 1. Chebyshev couldn't prove that, but he got close enough to obtain useful information.

Why are the zeros of the zeta function so important? A basic theorem in complex analysis states that subject to some technical conditions, a function of a complex variable is completely determined by the values of the variable for which that function is zero or infinity, together with some further information about its behaviour at those points. These special places are known as the function's zeros and poles. This theorem doesn't work in real analysis – one of many reasons why complex analysis became the preferred setting, despite requiring the square root of minus one. The zeta function has one pole, at $s = 1$, so everything about it is determined by its zeros provided we bear this single pole in mind.

For convenience, Riemann mostly worked with a related function, the xi function $\xi(x)$, which is closely related to the zeta function, and emerges from the method of analytic continuation. He remarked that:

> It is very probable that all [zeros of the xi function] are real. One would, however, wish for a strict proof of this; I have, though, after some fleeting futile attempts, provisionally put aside the search for such, as it appears unnecessary for the next objective of my investigation.

This statement about the xi function is equivalent to one about the related zeta function. Namely, all nontrivial zeros of the zeta function are complex numbers of the form $\frac{1}{2} + it$: they lie on the *critical line* 'real part equals $\frac{1}{2}$,' Figure 35. This version of his remark is the famous Riemann hypothesis.

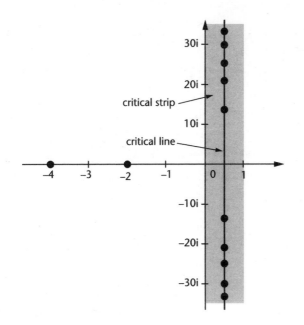

Fig 35 Zeros of the zeta function, the critical line, and the critical strip.

Riemann's remark is rather casual, as if the Riemann hypothesis isn't terribly important. For his programme to prove the prime number theorem, it wasn't. But for many other questions, the reverse is true. In fact, the Riemann hypothesis is widely considered to be the most important unanswered question in mathematics.

To understand why, we must pursue Riemann's thinking a little further. He had his sights on the prime number theorem. His exact formula suggested a way to achieve this: understand the zeros of the zeta function, or equivalently the xi function. The full Riemann hypothesis is not required; you just have to prove that all nontrivial zeros of the zeta function have real parts between 0 and 1. That is, they lie within distance $\frac{1}{2}$ of Riemann's critical line, in the so-called critical strip. This property of the zeros implies that the sum over the zeros of the zeta function, in the exact formula above, is a finite constant. Asymptotically, for large x, it might as well not be there at all. Among the terms in the formula, the only one that remains important as x becomes very large is x. All of the complicated stuff disappears asymptotically by comparison with x. Therefore the weighted count is asymptotic to x, and that proves the prime number theorem. So, ironically, the role of the zeros of the zeta function is to prove that the zeros of the zeta function do not make a significant contribution to the exact formula.

Riemann never pushed this programme through to a conclusion. In fact, he never again wrote on the topic. But two other mathematicians took up the challenge, and showed that Riemann's hunch was correct. In 1896, Jacques Hadamard and Charles Jean de la Vallée Poussin independently deduced the prime number theorem by proving that all nontrivial zeros of the zeta function lie in the critical strip. Their proofs were very complicated and technical; nonetheless, they worked. A new and powerful area of mathematics came into being: analytic number theory. It had applications throughout number theory, solving old problems and revealing new patterns. Other mathematicians found simpler analytic proofs of the prime number theorem, and Atle Selberg and Paul Erdős discovered a very complicated proof that did not require complex analysis at all. But by then Riemann's idea had been used to prove innumerable important theorems, including approximations to many number-theoretic functions. So their new proof added an ironic footnote, but otherwise had little effect. In 1980

Donald Newman found a much simpler proof, using only one of the most basic results in complex analysis, known as Cauchy's theorem.

Although Riemann declared his hypothesis to be unnecessary for his immediate objectives, it turned out to be vital to many other questions in number theory. Before discussing the Riemann hypothesis, it is worth taking a look at some of the theorems that would follow if the hypothesis could be proved true.

One of the most important implications is the size of the error in the prime number theorem. The theorem states that for large x, the ratio of $\pi(x)$ to $\mathrm{Li}(x)$ gets closer and closer to 1. That is, the size of the difference between those two functions shrinks to zero, *relative to the size of x*. However, the actual difference can (and does) become larger and larger. It just does so at a slower rate than x itself does.[66] Computer experiments suggest that the size of the error is roughly proportional to $\sqrt{x}\log x$. If the Riemann hypothesis is true, this statement can be proved. In 1901 Helge von Koch proved that the Riemann hypothesis is logically equivalent to the estimate

$$|\pi(x) - \mathrm{Li}(x)| \leqslant \frac{1}{8\pi}\sqrt{x}\log x$$

for all $x \geqslant 2657$. Here the vertical bars $|\ |$ indicate the absolute value: the difference multiplied by ± 1 to make it positive. This formula provides the best possible bound on the difference between $\pi(x)$ and $\mathrm{Li}(x)$.

The Riemann hypothesis implies many estimates for other number-theoretic functions. For example, it is equivalent to the sum of the divisors of n being less than

$$e^{\gamma} n \log\log n$$

for all $n \geqslant 5040$, where $\gamma = 0.57721 \ldots$ is Euler's constant.[67] These facts may seem like oddities, but good estimates for important functions are vital to many applications, and most number theorists would give their right arm to prove any of them.

The Riemann hypothesis also tells us how big the gaps between consecutive primes can be. We can deduce the typical size of this gap from the prime number theorem: on average, the gap between a prime

p and the next one is comparable to $\log p$. Some gaps are smaller, some bigger, and it would make mathematicians' lives easier if they knew how large the biggest gaps can become. Harald Cramér proved in 1936 that if the Riemann hypothesis is true, the gap at prime p is no bigger than a constant times $\sqrt{p}\log p$.

The true significance of the Riemann hypothesis lies much deeper. There are far-reaching generalisations, plus a strong hunch that anyone who can prove the Riemann hypothesis can probably prove the corresponding generalised Riemann hypothesis. Which in turn would give mathematicians a lot of control over wide-ranging areas of number theory.

The generalised Riemann hypothesis arises from a finer description of prime numbers. All primes other than 2 are odd, and we saw in chapter 2 that the odd ones can be classified into two kinds: those that are 1 greater than a multiple of 4, and those that are 3 greater than a multiple of 4. They are said to be of the form $4k + 1$ or $4k + 3$, where k is what you multiply by 4 to get them. Here's a short list of the first few primes of each type, together with the corresponding multiples of 4:

multiple of 4	0	4	8	12	16	20	24	28	32	36
plus 1	0	5	•	13	17	•	•	29	•	37
plus 3	3	7	11	•	19	23	•	•	•	•

The dot indicates that the number concerned is not prime.

How many primes of each kind are there? How are they distributed among the primes, or among all integers? Euclid's proof that there are infinitely many primes can be modified, without much effort, to prove that there are infinitely many primes of the form $4k + 3$. It is much harder to prove that there are infinitely many primes of the form $4k + 1$; it can be done, but only by using some fairly difficult theorems. The difference arises because any number of the form $4k + 3$ has some factor of that form; the same is not always true of numbers $4k + 1$.

There's nothing sacred about the numbers here. Apart from 2 and 3, all primes are either of the form $6k + 1$ or $6k + 5$, and we can ask similar questions. For that matter, all primes except 5 take one of the forms $5k + 1$, $5k + 2$, $5k + 3$, $5k + 4$. We leave out $5k$ because these are the multiples of 5, so all of them except 5 are not prime.

It's not hard to come up with a sensible guess for all questions of this kind – primes in an arithmetic sequence. The $5k$ case is typical. Experiment quickly suggests that numbers of the four types listed above have much the same chance of being prime. Here's a similar table:

multiple of 5	5	10	15	20	25	30	35	40
plus 1	•	11	•	•	•	31	•	41
plus 2	7	•	17	•	•	•	37	•
plus 3	•	13	•	23	•	•	•	43
plus 4	•	•	19	•	29	•	•	•

So there should be infinitely many of each individual type, and on average about one quarter of the primes, up to some given limit, should be of any specific form.

Simple proofs show that some forms lead to infinitely many primes, more sophisticated proofs work for other forms, but until the mid-1800s no one could prove there are infinitely many primes of each possible form, let alone that the proportions are roughly equal. Lagrange assumed this without proof in his work on the law of quadratic reciprocity – a deep property of squares to a prime modulus – in 1785. The results clearly had useful consequences, and it was high time someone proved them. In 1837 Dirichlet discovered how to adapt Riemann's ideas about the prime number theorem to prove both of those statements. The first step was to define analogues of the zeta function for these types of prime. The resulting functions are called Dirichlet L-functions. An example, arising in the $4k + 1/4k + 3$ case, is

$$L(s, \chi) = 1 - 3^{-s} + 5^{-s} - 7^{-s} + 9^{-s} - \cdots$$

where the coefficients are $+1$ for numbers of the form $4k + 1$, -1 for numbers $4k + 3$, and 0 for the rest. The Greek letter χ is called a Dirichlet character, and it reminds us to use these signs.

For Riemann's zeta function what matters is not just the series but its analytic continuation, which gives the function a meaning for all complex numbers. The same goes for the L-function, and Dirichlet defined a suitable analytic continuation. By adapting the ideas used to prove the prime number theorem, he was then able to prove an analogous theorem for primes of specific forms. For example, the number of primes of the form $5k + 1$ less than or equal to x is asymptotic to $\mathrm{Li}(x)/4$, and the same goes for the other three cases $5k + 2$, $5k + 3$, $5k + 4$. In particular there are infinitely many primes of each form.

The Riemann zeta function is a special case of a Dirichlet L-function for primes of the form $1k + 0$, that is, all primes. The generalised Riemann hypothesis is the obvious generalisation of the original Riemann hypothesis: the zeros of any Dirichlet L-function either have real part $\frac{1}{2}$, or they are 'trivial zeros' with real part either negative or greater than 1.

If the generalised Riemann hypothesis is true, then so is the Riemann hypothesis. Many of the consequences of the generalised Riemann hypothesis are analogues of those for the Riemann hypothesis. For example, similar error bounds can be proved for the analogous versions of the prime number theorem, applied to primes of any specific form. However, the generalised Riemann hypothesis implies many things that are quite different from anything that we can derive using the ordinary Riemann hypothesis. Thus in 1917 Godfrey Harold Hardy and John Edensor Littlewood proved that the generalised Riemann hypothesis implies a conjecture of Chebyshev, to the effect that (in a precise sense) primes of the form $4k + 3$ are more common than those of the form $4k + 1$. Both types are equally likely, in the long run, by Dirichlet's theorem, but that doesn't stop the $4k + 3$ primes outcompeting the $4k + 1$ primes if you set up the right game.

The generalised Riemann hypothesis also has important implications for primality tests, such as Miller's 1976 test mentioned in chapter 2. If the generalised Riemann hypothesis is true, then Miller's test provides an efficient algorithm. Estimates of the efficiency of more recent tests also depend on the generalised Riemann hypothesis. There are significant applications to algebraic number theory, too. Recall from chapter 7 that Dedekind's reformulation of Kummer's ideal numbers led to a new and fundamental concept, ideals. Prime factorisation in rings of algebraic integers exists, but may not be unique. Prime factorisation of ideals is much tidier: both existence and uniqueness are valid. So it makes sense to reinterpret all questions about factors in terms of ideals. In particular, there is a notion of a 'prime ideal', a reasonable and tractable analogue of a prime number.

Knowing this, it is natural to ask whether Euler's link between ordinary primes and the zeta function has an analogue for prime ideals. If so, the whole powerful machinery of analytic number theory becomes available for algebraic numbers. It turns out that this can be done, with deep and vital implications. The result is the Dedekind zeta function: one such function for each system of algebraic numbers. There is a deep link between the complex analytic properties of the Dedekind zeta function and the arithmetic of prime numbers for the corresponding algebraic integers. And, of course, there is an analogue of the Riemann hypothesis: all nontrivial zeros of the Dedekind zeta function lie on the critical line. The phrase 'generalised Riemann hypothesis' now includes this conjecture as well.

Even this generalisation is not the end of the story of the zeta function. It has inspired the definition of analogous functions in several other areas of mathematics, from abstract algebra to dynamical systems theory. In all of these areas there are even more far-reaching analogues of the Riemann hypothesis. A few of them have even been proved to be true. In 1974 Pierre Deligne proved such an analogue for varieties over finite fields. Generalisations known as Selberg zeta functions satisfy an analogue of the Riemann hypothesis. The same goes for the Goss zeta function. However, there exist other generalisations, Epstein zeta functions, for which the appropriate analogue of the Riemann hypothesis is false. Here infinitely many nontrivial zeros lie on the critical line, but some do not, as Edward Titchmarsh demonstrated in 1986. On the other hand, these zeta functions do not have an Euler product formula, so they fail to resemble the Riemann zeta functions in what may well be a crucial respect.

The circumstantial evidence in favour of the truth of the Riemann hypothesis – either the original, or its generalisations – is extensive. Many beautiful things would follow from the truth of the hypothesis. None of these things has ever been disproved: to do so would be to disprove the Riemann hypothesis, but neither proof nor disproof is known. There is a widespread feeling that a proof of the original Riemann hypothesis would open the way to a proof of its generalisations as well. In fact, it might be better to attack the generalised Riemann hypothesis in all its glory, exploiting the wealth of methods now available, and then to deduce the original Riemann hypothesis as a special case.

There is also a vast amount of experimental evidence for the truth of the Riemann hypothesis – or what certainly looks like a vast amount until someone throws cold water on that claim. According to Carl Ludwig Siegel, Riemann calculated the first few zeros of his zeta function numerically but did not publish the results: they are located at

$$\tfrac{1}{2} \pm 14.135i \qquad \tfrac{1}{2} \pm 21.022i \qquad \tfrac{1}{2} \pm 25.011i$$

The nontrivial zeros always come in $\pm$ pairs like this. I've written $\tfrac{1}{2}$ here rather than 0.5 because the real part is known *exactly* in these cases, by exploiting general results in complex analysis and known properties of the zeta function. The same goes for the computer calculations reported below. They don't just show that zeros are very close to the critical line; they are actually on it.

In 1903 Jorgen Gram showed numerically that the first ten ($\pm$ pairs of) zeros lie on the critical line. By 1935 Titchmarsh had increased the number to 195. In 1936 Titchmarsh and Leslie Comrie proved that the first 1041 pairs of zeros are on the critical line – the last time anyone did such computations by hand. Alan Turing is best known for his wartime efforts at Bletchley Park, where he helped to break the German Enigma code, and for his work on the foundations of computing and artificial intelligence. But he also took an interest in analytic number theory. In 1953 he discovered a more efficient method for calculating zeros of the zeta function, and used a computer to deduce that the first 1104 pairs of zeros are on the critical line. Evidence for all zeros up to some limit being on the critical line piled up; the current record, obtained by Yannick Saouter and Patrick Demichel in 2004, is 10 trillion (10^{13}). Various mathematicians and computer scientists have also checked other ranges of zeros. To date, every nontrivial zero that has been computed lies on the critical line.

This might seem conclusive, but mathematicians are ambivalent about this kind of evidence, with good reason. Numbers like 10 trillion may sound large, but in number theory what often matters is the logarithm of the number, which is proportional to the number of digits. The logarithm of 10 trillion is just under 30. In fact, many problems hinge on the logarithm of the logarithm, or even the logarithm of the logarithm of the logarithm. In those terms, 10 trillion is *tiny*, so numerical evidence up to 10 trillion carries hardly any weight.

There is also some general analytic evidence, which is not subject to this objection. Hardy and Littlewood proved that infinitely many zeros lie on the critical line. Other mathematicians have shown, in a precise sense, that almost all zeros lie very close to the critical line. Selberg proved that a nonzero proportion of zeros lie on the critical line. Norman Levinson proved that this proportion is at least one third, a figure now improved to at least 40 per cent. All of these results suggest that if the Riemann hypothesis is false, zeros that do not lie on the critical line are very large, and very rare. Unfortunately, the main implication is that if such exceptions exist, finding them will be extraordinarily hard.

Why bother? Surely the numerical evidence ought to satisfy any sensible person? Unfortunately not. It doesn't satisfy mathematicians, and in this instance they are not just being pedantic: they are indeed acting as sensible people. In mathematics in general, and especially in number theory, apparently extensive 'experimental' evidence often carries much less weight than you might imagine.

An object lesson is provided by the Pólya conjecture, stated in 1919 by the Hungarian mathematician George Pólya. He suggested that at least half of all whole numbers up to any specific value have an odd number of prime factors. Here repeated factors are counted separately, and we start from 2. For example, up to 20 the number of prime factors looks like Table 2, where the column 'percentage' gives the percentage of numbers up to this size with an odd number of prime factors.

All the percentages in the final column are bigger than 50 per cent, and more extensive calculations make it reasonable to conjecture that this is always true. In 1919, with no computers available, experiments could not find any numbers that disproved the conjecture. But in 1958

number	factorisation	how many primes?	percentage
2	2	1	100
3	3	1	100
4	2^2	2	66
5	5	1	75
6	2×3	2	60
7	7	1	66
8	2^3	3	71
9	3^2	2	62
10	2×5	2	55
11	11	1	60
12	$2^2 \times 3$	3	63
13	13	1	66
14	2×7	2	61
15	3×5	2	57
16	2^4	4	60
17	17	1	62
18	2×3^2	3	64
19	19	1	66
20	$2^2 \times 5$	3	65

Table 2 Percentages of numbers, up to a given size, that have an odd number of prime factors.

Brian Haselgrove used analytic number theory to prove that the conjecture is false for some number – less than 1.845×10^{361}, to be precise. Once computers arrived on the scene, Sherman Lehman showed that the conjecture is false for 906,180,359. By 1980 Minoru Tanaka has proved that the smallest such example is 906,150,257. So you could have amassed experimental evidence in the conjecture's favour for nearly all numbers up to a billion, even though it is false.

Still, it's nice to know that the number 906,150,257 is unusually interesting.

Of course, today's computers would disprove the conjecture in a few seconds, if suitably programmed. But sometimes even computers don't help. A classic example is Skewes's number, where apparently huge amounts of numerical evidence initially suggested that a famous conjecture should be true, but in fact it is false. This gigantic number appeared in a problem closely related to the Riemann hypothesis: the approximation of $\pi(x)$ by $\text{Li}(x)$. As we've just seen, the prime number theorem states that the ratio of these two quantities tends to 1 as x becomes large. Numerical calculations seem to indicate something stronger: the ratio is always less than 1; that is, $\pi(x)$ is less than $\text{Li}(x)$. In 2008 Tadej Kotnik's numerical computations showed that this is true whenever x is less than 10^{14}. By 2012 Douglas Stoll and Demichel had improved this bound to 10^{18}, a figure obtained independently by Andry Kulsha. Results of Tomás Oliveira e Silva suggest it can be increased to 10^{20}.

This might sound definitive. It's stronger than the best numerical results we have for the Riemann hypothesis. But in 1914 Littlewood proved that this conjecture is false – spectacularly so. As x runs through the positive real numbers, the difference $\pi(x)-\text{Li}(x)$ changes sign (from negative to positive or the reverse) *infinitely often*. In particular, $\pi(x)$ is *bigger* than $\text{Li}(x)$ for some sufficiently large values of x. However, Littlewood's proof gave no indication of the size of such a value.

In 1933 his student, the South African mathematician Stanley Skewes, estimated how big x must be: no more than 10^10^10^34, where ^ indicates 'raised to the power'. That number is so gigantic that if all of its digits were printed in a book – rather a boring book, consisting of 1 followed by endless 0s – the universe would not be big enough to contain it, even if every digit were the size of a subatomic particle. Moreover, Skewes had to assume the truth of the Riemann hypothesis to make his proof work. By 1955 he had found a way to avoid the Riemann hypothesis, but at a price: his estimate increased to 10^10^10^963.

These numbers are too big even for the adjective 'astronomical', but further research reduced them to something that might be termed cosmological. In 1966 Lehman replaced Skewes's numbers by 10^{1165}. Te Riele reduced this to 7×10^{370} in 1987, and in 2000 Carter Bays and Richard Hudson reduced it to 1.39822×10^{316}. Kuok Fai Chow and Roger Plymen chipped a bit off, and got the number down to 1.39801×10^{316}. This may seem a negligible improvement, but it's about 2×10^{313} smaller. Saouter and Demichel made a further improvement to $1.3971667 \times 10^{316}$. Meanwhile in 1941 Aurel Wintner had proved that a small but nonzero proportion of integers satisfy $\pi(x) > \text{Li}(x)$. In 2011 Stoll and Demichel computed the first 200 billion zeros of the zeta function, which gives control of $\pi(x)$ when x is anything up to $10^{10,000,000,000,000}$, and found evidence that if x is less than 3.17×10^{114} then $\pi(x)$ is smaller than $\text{Li}(x)$.[68] So for this particular problem, the evidence up to at least 10^{18}, and very possibly up to 10^{114} or more, is completely misleading. The fickle gods of number theory are having an amusing joke at human expense.

Over the years, many attempts have been made to prove or disprove the Riemann hypothesis. Matthew Watkins's website 'Proposed proofs of the Riemann hypothesis' lists around 50 of them since 2000.[69] Mistakes have been found in many of these attempts, and none has been accepted as correct by qualified experts.

One of the most widely publicised efforts, in recent years, was that of Louis de Branges in 2002. He circulated a lengthy manuscript claiming to deduce the Riemann hypothesis by applying a branch of analysis that dealt with operators on infinite-dimensional spaces, known as functional analysis. There were reasons to take de Branges seriously. He had previously circulated a proof of the Bieberbach conjecture, about series expansions of complex functions. Although his original proof had errors, it was eventually established that the underlying idea worked. However, there now seem to be good reasons to think that de Branges's proposed method for proving the Riemann hypothesis has no chance of succeeding. Some apparently fatal obstacles have been pointed out by Brian Conrey and Xian-Jin Li.[70]

Perhaps the greatest hope for a proof comes from new or radically different ways to think about the problem. As we've seen repeatedly, breakthroughs on great problems often arise when someone links them to some totally different area of mathematics. Fermat's last theorem is a clear example: once it was reinterpreted as a question about elliptic curves, progress was rapid.

De Branges's tactics now seem questionable, but his approach is strategically sound. It has its roots in a verbal suggestion made around 1912 by David Hilbert, and independently by George Pólya. The physicist Edmund Landau asked Pólya for a physical reason why the

Riemann hypothesis should be true. Pólya related in 1982 that he had come up with an answer: the zeros of the zeta function should be related to the eigenvalues of a so-called self-adjoint operator. These are characteristic numbers associated with special kinds of transformation. In quantum physics, one of the important applications, these numbers determine the energy levels of the system concerned, and a standard and easy theorem states that the eigenvalues of this special kind of operator are always real. As we have seen, the Riemann hypothesis can be rephrased as the statement that all zeros of the xi function are real. If some self-adjoint operator had eigenvalues that were the same as the zeros of the xi function, the Riemann hypothesis would be an easy consequence. Pólya didn't publish this idea – he couldn't write down such an operator, and until someone could, it was pie in the sky. But in 1950 Selberg proved his 'trace formula', which relates the geometry of a surface to the eigenvalues of an associated operator. This made the idea a little more plausible.

In 1972 Hugh Montgomery was visiting the Institute for Advanced Study in Princeton. He had noticed some surprising statistical features of the nontrivial zeros of the zeta function. He mentioned them to the physicist Freeman Dyson, who immediately spotted a similarity to statistical features of random Hermitian matrices, another special type of operator used to describe quantum systems such as atomic nuclei. In 1999 Alain Connes came up with a trace formula, similar to Selberg's, whose validity would imply the truth of the generalised Riemann hypothesis. And in 1999 physicists Michael Berry and Jon Keating suggested that the required operator might arise by quantising a well-known concept from classical physics, related to momentum. The resulting Berry conjecture can be viewed as a more specific version of the Hilbert-Pólya conjecture.

These ideas, relating the Riemann hypothesis to core areas of mathematical physics, are remarkable. They show that progress may eventually come from apparently unrelated areas of mathematics, and raise hopes that the Riemann hypothesis may one day be settled. However, they have not yet led to any definitive breakthrough that encourages us to think that a solution is just around the corner. The Riemann hypothesis remains one of the most baffling and irritating enigmas in the whole of mathematics.

Today there is a new reason to try to prove the Riemann hypothesis: a substantial prize.

There is no Nobel prize in mathematics. The most distinguished prize in mathematics is the Fields medal, more properly the International Medal for Outstanding Discoveries in Mathematics. It is named after the Canadian mathematician John Fields, who endowed the award in his will. Every four years, at the International Congress of Mathematicians, up to four of the world's leading young (under 40) mathematicians receive a gold medal and a cash award, currently $15,000. As far as mathematicians are concerned, the Fields medal is equivalent in prestige to a Nobel prize.

Many mathematicians consider the lack of a Nobel in their subject to be a good thing. A Nobel prize is currently worth just over a million dollars, an amount that could easily distort research objectives and lead to arguments about priority. However, the absence of a major mathematical prize may also have distorted the public perception of the value and utility of mathematics. It's easy to imagine that if no one is willing to pay for it, it can't be worth much.

Recently two new high-prestige mathematical awards have come into being. One is the Abel prize, awarded annually by the Norwegian Academy of Science and Letters, and named after the great Norwegian mathematician Niels Henrik Abel. The other new award consists of seven Clay Mathematics Institute millennium prizes. The Clay Institute was founded by Landon Clay and his wife Lavinia. Landon Clay is an American businessman active in mutual funds, with a love of, and respect for, mathematics. In 1999 he established a new foundation for mathematics in Cambridge, Massachusetts, which runs meetings, awards research grants, organises public lectures, and administers an annual research award.

In 2000 Sir Michael Atiyah and John Tate, leading mathematicians in Britain and the United States, announced that the Clay Mathematics Institute had set up a new award, intended to encourage the solution of seven of the most important open problems in mathematics. They would be known as the millennium problems, and a properly published and refereed solution of any one of them would be worth 1 million dollars. Collectively, these problems draw attention to some of the central unanswered questions in mathematics, carefully selected by some of the world's top mathematicians. The substantial prize makes a very clear point to the public: mathematics is valuable. Everyone involved is aware that its intellectual value may be deeper than mere money, but a cash prize does help to concentrate minds. The best-known millennium problem, and the one that goes back furthest historically, is the Riemann hypothesis. It is the only question to feature both in Hilbert's list of 1900 and the list of millennium problems. The other six millennium problems are discussed in chapters 10–15. Mathematicians are not especially obsessed with prizes, and they would work on the Riemann hypothesis without one. A new and promising idea would be all the motivation they would need.

It's worth remembering that conjectures, however time-honoured, may not be true. Today, most mathematicians seem to think that a proof of the Riemann hypothesis will eventually be found. A few, however, think it may be false: somewhere out in the wilderness of very big numbers there may lurk a zero that does not sit on the critical line. If such a 'counterexample' exists it is likely to be very, very big.

However, opinions count for little at the frontiers of mathematics. Expert intuition is often very good indeed, but there have been plenty of occasions when it was wrong. The conventional wisdom can be both conventional and wise, without being true. Littlewood, one of the great experts in complex analysis, was unequivocal: in 1962 he said that he was sure the Riemann hypothesis was false, adding that there was no imaginable reason for it to be true. Who is right? We can only wait and see.

10

What shape is a sphere?
Poincaré Conjecture

HENRI POINCARÉ WAS ONE OF the greatest mathematicians of the late nineteenth century, a bit of an eccentric, but an astute operator. He became a member of France's Bureau des Longitudes, whose job was to improve navigation, time-keeping, and measurement of the Earth and planets. This appointment led him to propose setting up international time zones; it also inspired him to think about the physics of time, anticipating some of Einstein's discoveries in special relativity. Poincaré left his mark all over the mathematical landscape, from number theory to mathematical physics.

In particular, he was one of the founders of topology, the mathematics of continuous transformations. Here, in 1904, he ran into an apparently simple question, having belatedly realised that he had tacitly assumed the answer in earlier work, but couldn't find a proof. 'This question would lead us too far astray,' he wrote, which rather slid round the real issue: it wasn't leading him *anywhere*. Although he phrased the problem as a question, it became known as the Poincaré conjecture because everyone expected the answer to be 'yes'. It is another of the seven Clay millennium prize problems, and rightly so, because it turned out to be one of the most baffling problems in the whole of topology. Poincaré's question was finally answered in 2002 by a young Russian, Grigori Perelman. The solution introduced a whole raft of new ideas and methods, so much so that it took the mathematical community a few years to digest the proof and accept that it was correct.

For his success, Perelman was awarded a Fields medal, the most prestigious mathematical prize, but he declined it. He didn't want publicity. He was offered the million-dollar Clay prize for proving the Poincaré conjecture, and turned it down. He didn't want money, either. What he had wanted was for his work to be accepted by the mathematical community. Eventually it was, but unfortunately, for sensible reasons, that took a while. And it was always unrealistic to expect acceptance without publicity or the offer of prizes. But these unavoidable consequences of success didn't suit Perelman's sometimes reclusive nature.

We encountered topology in connection with the four colour theorem, and I resorted to the cliché 'rubber-sheet geometry'. Euclid's geometry deals with straight lines, circles, lengths, and angles. It takes place in a plane, or in a space of 3 dimensions when it becomes more advanced. A plane is like an infinite sheet of paper, and it shares one basic feature of paper: it doesn't stretch, shrink, or bend. You can roll paper into a tube, and it can shrink or stretch a tiny bit, especially if you spill coffee on it. But you can't wrap a sheet of paper round a sphere without creating creases. Mathematically, the Euclidean plane is rigid. In Euclid's geometry, two objects – triangles, squares, circles – are the same if you can transform one of them into the other by a rigid motion. And 'rigid' means that distances don't change.

What if you use an elastic sheet instead of paper? That does stretch, it does bend, and with a bit of effort it can be compressed. Lengths and angles have no fixed meaning on a sheet of elastic. Indeed, if it is elastic *enough*, neither do triangles, squares, or circles. You can deform a triangle on a sheet of rubber to give it an extra corner. You can even turn it into a circle, Figure 36. Whatever the concepts of rubber-sheet geometry are, they don't include the traditional Euclidean ones.

It might seem that geometry on a sheet of rubber would be so flexible that nothing would have a fixed meaning, in which case little of substance could be proved. Not so. Draw a triangle and place a point inside it. If you stretch and deform the sheet until the triangle becomes

Fig 36 Topological deformation of a triangle to a circle.

a circle, one feature of your diagram doesn't change: the point remains on the inside. Agreed, it's now inside a circle, not a triangle, but it's not *outside*. In order to move the point to the outside, you have to tear the sheet. That breaks the rules of this particular game.

There's another feature that survives distortion, too. A triangle is a simple closed curve. It is a line that joins up with itself so that there are no free ends, and does not cross itself. A figure-eight is a closed curve, but it's not simple – it crosses itself. When you deform the rubber sheet, the triangle may change shape, but it always remains a simple closed curve. There is no way to turn it into a figure-eight, for example, without tearing the sheet.

In three-dimensional topology, the whole of space becomes elastic. Not like a block of rubber, which twangs back into its original shape if you let go, but like a gel that can be deformed without any resistance. A topological space is infinitely deformable; you can take a region the size of a grain of rice and blow it up to the size of the Sun. You can pull out tentacles until the region is shaped like an octopus. The one thing you are not allowed to do is to introduce any kind of discontinuity. You mustn't tear space, or perform any kind of distortion that rips nearby points apart.

What features of a shape in space survive all continuous deformations? Not length, area, or volume. But being knotted does. If you tie a knot in a curve and join the ends to make a loop, then the knot can never escape. However you deform the space, the curve remains knotted. So we are working with a new kind of geometry in which the important and meaningful concepts seem rather fuzzy: 'inside', 'closed', 'simple', 'knotted'. This new geometry has a respectable name: topology. It may seem rather esoteric, perhaps even absurd, but it has turned out to be one of the major areas of twentieth-century mathematics, and it remains equally vital in the twenty-first. And one of the main people we have to thank for that is Poincaré.

The story of topology began to take off almost a century before Poincaré, in 1813. Simon Antoine Jean Lhuilier, a Swiss mathematician, didn't exactly set the world of mathematics alight during his lifetime, even though he turned down a large sum of money that a relative had promised to pay him if he entered the Church. Lhuilier preferred a career in mathematics. He specialised in a mathematical backwater: Euler's theorem for polyhedrons. In chapter 4 we came across this curious, apparently isolated result: if a polyhedron has F faces, V vertexes, and E edges, then $F - E + V = 2$. Lhuilier spent much of his career investigating variants of this formula, and in retrospect he took a vital step in the direction of topology when he discovered that Euler's formula is sometimes wrong. Its validity depends on the qualitative shape of the polyhedron.

The formula is correct for polyhedrons without any holes, which can be drawn on the surface of a sphere or continuously deformed into a shape of that kind. But when the polyhedron has holes, the formula fails. A picture-frame made from wood with a rectangular cross-section, for example, has 16 faces, 32 edges, and 16 vertexes; here $F - E + V = 0$. Lhuilier modified Euler's formula to cover these more exotic polyhedrons: if there are g holes, then $F - E + V = 2 - 2g$. This was the first discovery of a significant topological invariant: a quantity associated with a space, which does not change when the space is deformed continuously. Lhuilier's invariant provides a rigorous way to count how many holes a surface has, without the need to define 'hole'. This is useful, because the concept of a hole is tricky. A hole is not part of the surface, nor is it the region outside the surface. It appears to be a feature of how the surface sits in its surrounding space. But Lhuilier's discovery shows that what we interpret as the *number* of holes is an intrinsic feature, independent of any surrounding space. It is not necessary to define holes and then count them; in fact, it's better not to.

After Lhuilier, the next key figure in the prehistory of topology is Gauss. He encountered several other topological invariants when working on various core areas of mathematics. His work in complex analysis, especially the proof that every polynomial equation has at least one solution in complex numbers, led him to consider the winding number of a curve in the plane: how many times it winds round a given point. Problems in electricity and magnetism led to the linking number of two closed curves: how many times one of them winds through the other. These and other examples led Gauss to wonder whether there might exist some as-yet undiscovered branch of mathematics that would provide a systematic way to understand qualitative features of geometrical figures. He published nothing on the topic, but he mentioned it in letters and manuscripts.

He also passed the idea on to his student Johann Listing and his assistant August Möbius. I've mentioned the Möbius band, a surface with only one side and also one edge, which he published in 1865, and it can be found in Figure 9 of chapter 4. Möbius pointed out that 'having only one side', while intuitively clear, is difficult to make precise, and proposed a related property that could be defined in complete rigour. This property was orientability. A surface is orientable if you can cover it with a network of triangles, with arrows circulating round each triangle, so that whenever two triangles have a common edge the arrows point in opposite directions. If you draw a network on a plane and make all arrows run clockwise, for example, this is what happens. On a Möbius band, no such network exists.

Listing's first publication in topology came earlier, in 1847. Its title was *Vorstudien zur Topologie* ('Lectures on topology'), and it was the first text to employ that word. He had been using the term informally for about a decade. Another term used at that time is the Latin phrase *analysis situs*, 'analysis of position', but this eventually fell out of favour. Listing's book contains little of great significance, but it does set up a basic notion: covering a surface with a network of triangles. In 1861, four years ahead of Möbius, he described the Möbius band and studied connectivity: whether a space can be split into two or more disconnected parts. Building on Listing's work, a number of mathematicians, among them Walther von Dyck, put together a complete topological classification of surfaces, assuming them to be closed (no edge) and compact (of finite extent). The answer is that every orientable surface is topologically equivalent to a sphere, to which a finite number g of handles has been attached, Figure 11 (middle and right) in chapter 4. The number g is called the genus of the surface, and it is what Lhuillier's invariant determines. If $g = 0$ we have the sphere, and if $g > 0$ we obtain a torus with g holes. A similar sequence of surfaces, starting with the simplest non-orientable surface, the projective plane, classifies all non-orientable surfaces. The method was extended to allow surfaces with edges as well. Each edge is a closed loop, and the only extra information needed is how many of these loops occur.

The Poincaré conjecture will make more sense if we first take a look at one of the basic techniques employed in classifying surfaces. Earlier, I described topology in terms of deforming a shape made of rubber or gel, and emphasised the need to use *continuous* transformations. Ironically, one of the central techniques in topology involves what at first sight seems to be a discontinuous transformation: cut the shape into pieces. However, continuity is restored by a series of rules, describing which piece is joined to which, and in what manner. An example is the way we defined a torus by identifying opposite edges of a square, Figure 12 of chapter 4.

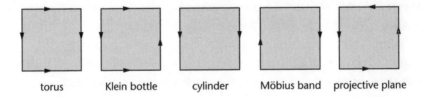

| torus | Klein bottle | cylinder | Möbius band | projective plane |

Fig 37 Five different topological spaces obtained by identifying opposite edges of a square in various ways.

Identifying points that appear to be distinct allows us to represent complicated topological spaces using simpler ingredients. A square is a square is a square, but a square with identification rules can be a torus, a Klein bottle, a cylinder, a Möbius band, or a projective plane, depending on the choice of rules, Figure 37. So when I explained a continuous transformation in terms of stretching and bending a rubber sheet, I asked for more than is strictly necessary. We are also permitted to cut the sheet, at an intermediate stage, provided that eventually we either join the edges together again exactly as they were to begin with, or we specify rules that have the same effect. As far as a topologist is concerned, *stating* a rule for gluing edges together is the same as actually implementing the rule. Provided you don't forget the rule in whatever else you do later.

The classic method for classifying surfaces begins by drawing a network of triangles on the surface. Then we cut enough of the edges to fold the triangles out flat to make a polygon. Gluing rules, derived from how we made the cuts, then specify how to identify various edges of the polygon, reconstructing the original surface. At that point, all of the interesting topology is implicit in the gluing rules. The classification is proved by manipulating the rules algebraically, and transforming them into rules that define a g-holed torus or one of the analogous non-orientable surfaces. Modern topology has other ways to achieve the same result, but it often uses this kind of 'cut and paste' construction. The method generalises without difficulty to spaces of any dimension, but it is too restricted to lead to a classification of higher-dimensional topological spaces without further assistance.

Around 1900, Poincaré was developing the earlier work on the topology of surfaces into a far more general technique, which applied to spaces with any number of dimensions. The main thrust of his research was to find topological invariants: numbers or algebraic formulas associated with spaces, which remain unchanged when the space is continuously deformed. If two spaces have different invariants, then one cannot be deformed into the other, so they are topologically distinct.

He started from the Italian mathematician Enrico Betti's generalisation of Lhuilier's topological invariant $F - E + V$, which is now rather unfairly known as the Euler characteristic, to higher-dimensional spaces, achieved in 1870. Betti had noticed that the largest number of closed curves that can be drawn on a surface of genus g, without dividing it into disconnected pieces, is $g - 1$. This is another way to characterise the surface topologically. He generalised this idea to 'connectivity numbers' of any dimension, which Poincaré called

Betti numbers, a term still in use today. The k-dimensional Betti number counts the number of k-dimensional holes in the space.

Poincaré developed Betti's connectivity numbers into a more sensitive invariant called homology, which has a lot more algebraic structure. We will discuss homology in more detail in chapter 15. Suffice it to say that it looks at collections of multidimensional 'faces' in this kind of network, and asks which of them form the boundary of a topological disc. A disc has no holes, unlike a torus, so we can be sure that within any collection of faces that constitutes a boundary, there are no holes. Conversely, we can detect holes by playing off collections of faces that do not form boundaries against collections that do. In this manner we can construct a series of invariants of a space, known as its homology groups. 'Group' here is a term from abstract algebra; it means that any two objects in the group can be combined to give something else in the same group, in a way that is subject to several nice algebraic rules. I'll say a bit more later when we need this idea. There is one such group for each dimension from 0 to n, and for each space we get a series of topological invariants, with all sorts of fascinating algebraic properties.

Listing had classified all topological surfaces – spaces of dimension 2. The obvious next step was to look at dimension 3. And the simplest space to start with was a sphere. In everyday language the word 'sphere' has two different meanings: it can be a solid round ball, or just the surface of the ball. When working on the topology of surfaces, the word 'sphere' is always interpreted in the second sense: the infinitely thin surface of a ball. Moreover, the inside of the sphere isn't considered to be part of it: it is just a consequence of the usual way we embed a spherical surface in space. Intrinsically, all we have is a surface, topologically equivalent to the surface of a ball. You can think of the sphere as a hollow ball with an infinitely thin skin.

The 'correct' 3-dimensional analogue of a sphere, called a 3-sphere, is *not* a solid ball. A solid ball is 3-dimensional, but it has a boundary: its surface, the sphere. A sphere doesn't have a boundary, and neither should its 3-dimensional analogue. The simplest way to define a 3-sphere is to mimic the coordinate geometry of an ordinary sphere. This leads to a space that is a little tricky to visualise: I can't show you a model in 3 dimensions because the 3-sphere – even though it has only

3 dimensions – doesn't embed in ordinary 3-dimensional space. Instead, it embeds in 4-dimensional space.

The usual unit sphere, in 3-dimensional space, consists of all points that are distance 1 from a specified point: the centre. Analogously, the unit 3-sphere in 4-dimensional space consists of all points that are unit distance from the centre. In coordinates we can write down a formula for this set using a generalisation of Pythagoras's theorem to define distance.[71] More generally, a 3-sphere is *any* space that is topologically equivalent to the unit 3-sphere, just as all sorts of lumpy versions of a unit 2-sphere are topological 2-spheres, and of course the same goes in higher dimensions.

If you're not satisfied by that, and want a more geometric image, try this one. A 3-sphere can be represented as a solid ball whose entire surface is identified with a single point. This is another example of a gluing rule, and in this case it is analogous to a way to turn a circular disc into a 2-sphere. If you run a cord round the edge of a cloth disc and draw it tight, like closing a bag, the result is topologically the same as a 2-sphere. Now perform the analogous operation on a solid ball, but as usual, do not try to visualise the result: just think of a solid ball and implement the gluing rules conceptually.

Anyway, Poincaré was very interested in the 3-sphere, because it was presumably the simplest 3-dimensional topological space that had no boundary and was of finite extent. In 1900 he published a paper in which he claimed that homology groups were a sufficiently powerful invariant to characterise the 3-sphere topologically. Specifically, if a 3-dimensional topological space has the same homology groups as a 3-sphere, then it is topologically equivalent to (can be continuously deformed into) a 3-sphere. By 1904, however, he had discovered that this claim is wrong. There is at least one 3-dimensional space that is not a 3-sphere, but has the same homology groups as a 3-sphere. The space was a triumph for the gluing rules philosophy, and the proof that it was not a 3-sphere involved the creation of a new invariant, necessarily more powerful than homology.

First, the space. It is called Poincaré's dodecahedral space, because a modern construction uses a solid dodecahedron. Poincaré was unaware of its relation to a dodecahedron: he glued two solid toruses together in a very obscure manner. The dodecahedron interpretation was published in 1933, some 21 years after Poincaré's death, by

Herbert Seifert and Constantin Weber, and it is much easier to comprehend. The analogy to bear in mind is the construction of a torus by gluing together opposite edges of a square. As always, you don't try to *do* the gluing; you just remember that corresponding points are considered to be the same. Now we do the same thing, but using opposite faces of a dodecahedron, Figure 38.

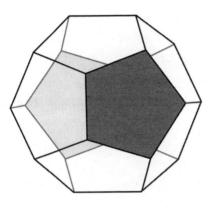

Fig 38 To make the Poincaré dodecahedral space, take a dodecahedron and glue all pairs of opposite faces (such as the shaded pair) together, with a twist to make them fit.

The Pythagoreans knew about dodecahedrons, 2500 years ago. The boundary of a dodecahedron consists of 12 regular pentagons, joined to make a roughly spherical cage, with three pentagons meeting at each corner. Now glue each face of the dodecahedron to the opposite face ... except that there's a twist. Literally. Each face has to be rotated through a suitable angle before it is glued to the opposite one. The angle is the smallest one that aligns the corresponding faces, which is 36 degrees. You can think of this rule as an elaborate version of the Möbius band rule: twist an edge through 180 degrees and then glue it to the opposite edge.

That's the space. Now let's look at the invariant. I'm not just wool-gathering: we need all this to understand the Poincaré conjecture.

Poincaré called his new invariant the fundamental group. We still use this name today, but we also refer to it as the (first) homotopy group. Homotopy is a geometrical construction that can be carried out entirely inside the space, and it provides information on the topological type of that space. It does that using an abstract algebraic structure known as a group. A group is a collection of mathematical objects, any two of which can be combined to give another object in the group. This law of combination – often called multiplication or addition, even when it's not the usual arithmetical operation with that name – is required to satisfy a few simple and natural conditions. If we call the operation addition, the main conditions are:

- The group contains an element that behaves like zero: if you add it to anything in the group, you get the same thing as a result.

- Every member of the group has a negative in the group: add the two and you get zero.

- If you add three members of the group together, it doesn't matter which two you add first. That is, $(a + b) + c = a + (b + c)$. This is called the associative law.

The one algebraic law that is *not* imposed (although sometimes it is true as well) is the commutative law $a + b = b + a$.[72]

Poincaré's fundamental group is a sort of simplified skeleton of the space. It is a topological invariant: topologically equivalent spaces have the same fundamental group. To gain useful insight, and very possibly to reconstruct part of Poincaré's motivation, let's see how it works for a circle, by stealing an image that goes back to Gauss. Imagine an ant whose entire universe is the circle. How can it find out what shape its universe is? Can it distinguish the circle from, say, a line? Bear in mind throughout that the ant is not permitted to step outside its universe, look at it, and see that it is circular. All it can do is wander around inside its universe, whatever that may be. In particular, the ant won't realise that its universe is bent, because its version of a light ray is also confined to the circle. Please ignore practicalities such as objects having to pass through each other – it's going to be a very loose analogy.

The ant can discover the shape of its universe in several ways. I'll focus on a method that generalises to any topological space. For the purposes of this discussion, the ant is a point. It lives at a bus stop, which is also a point. Every day it starts from home, takes the bus (which, of course, is a point) and ends up back home again. The most straightforward trip is the number 0 bus, which just sits at the stop and goes nowhere. For a more interesting excursion, the ant catches the number 1 bus, which goes round the universe exactly once in an anticlockwise direction and stops when it returns home. The number 2 bus goes round twice, the number 3 bus goes round three times, and so on: one anticlockwise bus for each positive integer. There are also negative buses, which go the other way. The number −1 bus goes round once clockwise, the number −2 bus goes round twice clockwise, and so on.

The ant quickly notices that two successive trips on the number 1 bus are essentially the same as a single trip on the number 2 bus, and three trips on the number 1 bus are essentially the same as a single trip on the number 3 bus. Similarly, a trip on the number 5 bus followed by a trip on the number 8 bus is essentially the same as a trip on the number 13 bus. In fact, given any two positive numbers, a trip on the bus with the first number, followed by a trip on the bus with the second number, boils down to a trip on the bus whose number is their sum.

The next step is more subtle. The same relationship *nearly* holds for buses whose numbers are negative or zero. A trip on the number 0 bus, followed by a trip on the number 1 bus, is very similar to a trip on the number 1 bus. However, there is a slight difference. On the $0 + 1$ trip, bus 0 waits around for a time at the start, which doesn't happen for a single trip on bus 1. So we introduce a notion with the forbidding name of homotopy ('same place' in Greek). Two loops are homotopic if one can be continuously deformed into the other. If we allow bus itineraries to be changed by homotopies, we can gradually shrink the time that the ant spends sitting at the bus stop on the number 0 bus, until the stationary period vanishes. Now the difference between the $0 + 1$ trip and the 1 trip has disappeared, so 'up to homotopy' the result is just a trip on the number 1 bus. That is, the bus number equation $0 + 1 = 1$ remains valid – not for trips, but for homotopy classes of trips.

What about a trip on the number 1 bus followed by a trip on the number −1 bus? We'd like this to be a trip on the number 0 bus, but it's not. It goes all the way round anticlockwise, and then comes back again all the way round clockwise. This is clearly different from spending the whole trip sitting at the bus stop on the number 0 bus. So $1 + (-1)$, that is, $1 - 1$, is not equal to 0. But again homotopy comes to the rescue: the combination of buses 1 and −1 is homotopic to the same overall trip as bus 0. To see why, suppose the ant follows the combined route for buses 1 and −1 by car, but just before it gets all the way round to the bus stop it reverses direction and goes home again. This trip is very close to the double bus trip: it just misses out one tiny part of the journey. So the original double bus trip has 'shrunk', continuously, to a slightly shorter car journey. Now the ant can shrink the journey again, by turning back slightly earlier. It can keep shrinking the journey, gradually turning back earlier and earlier, until eventually all it does is to sit in a parked car at the bus stop, going nowhere. This shrinking process is also a homotopy, and it shows that a trip on the number 1 bus followed by a trip on the number −1 bus is homotopic to a trip on the number 0 bus. That is, $1 + (-1) = 0$ for homotopy classes of trips.

It is now straightforward, for an algebraist, to prove that a trip on any bus, followed by a trip on a second bus, is homotopic to a trip on the bus you get by adding the two bus numbers. This is true for positive buses, negative buses, and the zero bus. So if we add bus trips together – well, homotopy classes of bus trips – we obtain a group. In fact, it's a very familiar group. Its elements are the integers (bus numbers) and its operation is addition. Its conventional symbol is $\mathbb{Z}$, from the German Zahl ('integer').

A lot more hard work proves that in a circular universe, *any* round trip by car – even if it involves lots of backtracking, reversing, going to and fro over the same stretch of road – is homotopic to one of the standard bus trips. Moreover, bus trips with different numbers are not homotopic. The proof requires some technique; the basic idea is Gauss's winding number. This counts the total number of times that the trip winds round the circle in the anticlockwise direction.[73] It tells you which bus route your journey is homotopic to.

Once the details are filled in, this description proves that the fundamental group of a circle is the same as the group $\mathbb{Z}$ of integers under addition. To add trips, just add their winding numbers. The ant could use this topological invariant to distinguish a circular universe from, say, an infinite line. On a line, any trip, however much it jiggles around, must at some stage reach a maximum distance from home. Now we can shrink the trip continuously by shrinking all distances from home by the same amount – first to 99 per cent, then 98 per cent, and so on. So on a line, *any* journey is homotopic to zero: staying at home. The fundamental group of the line has only one element: 0. Its algebraic properties are trivial: $0 + 0 = 0$. So it is called the trivial group, and since it differs from the group of all integers, the ant can tell the difference between living on a line and living on a circle.

As I said, there are other methods, but this is how the ant can do it using Poincaré's fundamental group.

Now we up the ante (pun intentional). Suppose that the ant lives on a surface. Again, that is its entire universe; it can't step outside and take a look to see what kind of surface it inhabits. Can it work out the topology of its universe? In particular, can it tell the difference between a sphere and a torus? Again the answer is 'yes', and the method is the same as for a circular universe: get on a bus and make round trips that start and finish from home. To add trips together, perform them in turn. The zero trip is 'stay at home', the inverse of a trip is the same trip in the opposite direction, and we get a group provided we work with homotopy classes of trips. This is the fundamental group of the surface. Compared to a circular universe there is more freedom to create trips and to deform them continuously into other trips, but the same basic idea works.

The fundamental group is again a topological invariant, and the ant can use it to find out whether it lives on a sphere or a torus. If its universe is a sphere, then no matter which trip the ant takes, it can gradually be deformed into the zero trip: stay at home. This is not the case if its universe is a torus. Some trips can be deformed to zero, but a trip that winds once through the central hole, as in Figure 39 (left), cannot be. That statement needs proof, but this can be provided. There are standard bus trips on the torus, but now the bus numbers are pairs of integers (m, n). The first number m specifies how many times the trip winds through the central hole; the second number n specifies how many times the trip winds round the torus. Figure 39 (right) shows the trip (5,2), which winds five times through the hole and twice round the torus. To add trips, add the corresponding numbers; for example, $(3, 6) + (2, 4) = (5, 10)$. The fundamental group of the torus is the group $\mathbb{Z}^2$ of *pairs* of integers.

Any topological space has a fundamental group, defined in exactly

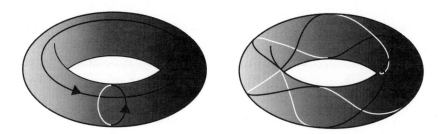

Fig 39 *Left:* Bus trips (1,0) and (0,1) on the torus. *Right:* Bus trip (5,2). Grey lines are at the back.

the same way using trips – more properly known as loops – that start and finish at the same point. Poincaré invented the fundamental group to prove that his dodecahedral space is not a 3-sphere, despite having exactly the same homology invariants. His original recipe is nicely adapted to the calculation of its fundamental group. The more modern 'twist and glue' recipe is even better adapted. The answer turns out to be a group with 120 elements related to the dodecahedron. In contrast, the fundamental group of a 3-sphere has only one element: the zero loop. So the dodecahedral space is *not* topologically equivalent to a sphere, despite having the same homology, and Poincaré had proved that his 1900 claim was wrong.

He went on to speculate about his new invariant: was that the missing ingredient in a topological characterisation of the 3-sphere? Perhaps any 3-dimensional space with the same fundamental group as a 3-sphere – that is, the trivial group – must actually *be* a 3-sphere? He phrased this suggestion in a negative way as a question: 'Consider a compact 3-dimensional manifold [topological space] V without boundary. Is it possible that the fundamental group of V could be trivial, even though V is not [topologically equivalent to] the 3-dimensional sphere?' He left the question open, but the very plausible belief that the answer is the obvious one – 'no', when the question is stated that way – quickly became known as the Poincaré conjecture. And it equally quickly became one of the most notorious open questions in topology.

'Trivial fundamental group' is another way to say 'every loop can be continuously deformed to a point'. Not only does a 3-sphere possess that property; so does an analogous n-sphere for any dimension n. So we can make the same conjecture for a sphere of any dimension. This statement is the n-dimensional Poincaré conjecture. It is true when $n = 2$, by the classification theorem for surfaces. And for over 50 years, that was as far as anyone could get.

In 1961 Stephen Smale borrowed a trick from the classification of surfaces, and applied it in higher dimensions. One way to think of a g-holed torus is to start with a sphere and add g handles – just like the handle on a teacup or a jug. Smale generalised this construction to any number of dimensions, calling the process handle decomposition. He analysed how handles could be modified without changing the topology of the space, and deduced the Poincaré conjecture in all dimensions greater than or equal to 7. His proof didn't work for lower dimensions, but other mathematicians found ways to repair it: John Stallings for dimension 6 and Christopher Zeeman for dimension 5. However, one vital step, known as the Whitney trick, failed for dimensions 3 and 4 because in these spaces there isn't enough room to perform the required manoeuvres, and no one could find effective substitutes. A general feeling emerged that topology, for spaces of these two dimensions, might be unusual.

This conventional wisdom was shaken in 1982 when Michael Freedman discovered a proof of the 4-dimensional Poincaré conjecture that did not require the Whitney trick. It was extremely complicated, but it worked. So, after 50 years with little progress and 20 years of frenzied activity, topologists had polished off the Poincaré conjecture in every dimension except the one Poincaré had originally asked about. The successes were impressive, but the methods used to obtain them provided very little insight into the 3-dimensional case. A different way of thinking was needed.

What finally broke the deadlock was rather like the traditional list of wedding gifts: something old, something new, something borrowed … and, stretching a point, something blue. The old idea was to revisit an area of topology which, after all of the flurry of activity on spaces of higher dimension, was generally thought to have been mined out: the topology of surfaces. The new idea was to rethink the classification of surfaces from a point of view that at first seemed completely alien:

classical geometry. The borrowed idea was the Ricci flow, which took its motivation from the mathematical formalism of Einstein's theory of general relativity. And the blue idea was 'blue sky' speculation: some far-reaching suggestions based on a dash of intuition and a lot of hope.

Recall that orientable surfaces without boundary can be listed: each is topologically equivalent to a torus with some number of holes. This number is the genus of the surface, and when it is zero, the surface is a sphere with no handles – that is, a sphere. The very word reminds us that among all topological spheres, there is one surface that stands out from all others as the archetype. Namely, the unit sphere in Euclidean space. Forget, for a second, all that rubber-sheet stuff. We'll put that back in a moment. Concentrate on the good old Euclidean sphere. It has all sorts of extra mathematical properties, coming from the rigidity of Euclid's geometry. Paramount among those properties is curvature. Curvature can be quantified; at each point on a geometric surface there is a number that measures how curved the surface is near that point. The sphere is the only closed surface in Euclidean space whose curvature is the same at every point, and is positive.

This is weird, because constant curvature is not a topological property. Even weirder: the sphere is not alone. There is also one standard geometric surface that stands out as the archetypal torus. Namely, start with a square in the plane, and identify opposite edges, Figure 12 of chapter 4. When we draw the result in 3-dimensional space, rolling up the square to make its edges meet, the result looks curved. But from an intrinsic point of view, we can work entirely with the square plus the gluing rules. A square has a natural geometric structure: it is a region in the Euclidean plane. The plane also has constant curvature, but now the constant is *zero*. A torus with this particular geometry also has zero curvature, so it is called the *flat torus*. The name may sound like an oxymoron, but for an ant living on a flat torus, carting a ruler and protractor around to measure lengths and angles, the local geometry would be identical to that of the plane.

The geometers of the eighteenth century, trying to understand Euclid's axiom about the existence of parallel lines, set out to deduce that axiom from the rest of Euclid's basic assumptions, failed repeatedly, and ended up realising that no such deduction is possible. There are three different kinds of geometry, each of which obeys every condition that Euclid requires, save for the parallel axiom. Those geometries are called Euclidean (the plane, where the parallel axiom is valid), elliptic (geometry on the surface of a sphere, with a few bells and whistles, where any two lines meet and parallels do not exist), and hyperbolic (where some lines fail to meet, and parallels fail to be unique). Moreover, the classical mathematicians interpreted these geometries as the geometry of curved spaces. Euclidean geometry corresponds to zero curvature, elliptic/spherical geometry corresponds to constant positive curvature, and hyperbolic geometry corresponds to constant negative curvature.

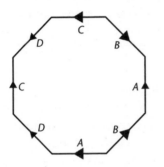

Fig 40 Making a 2-holed torus from an octagon by identifying edges in pairs (*AA, BB, CC, DD*).

We have just seen how to obtain the first two of these geometries: they occur on the sphere and the flat torus. In terms of the classification theorem, these are g-holed toruses for $g = 0$ and 1. The only thing missing is hyperbolic geometry. Does every g-holed torus have a natural geometric structure, based on taking some polygon in hyperbolic space and identifying some of its edges? The answer is striking: it is 'yes' for *any* value of g greater than or equal to 2. Figure 40 shows an example for $g = 2$ based on an octagon. I'll skip the hyperbolic geometry and the identification of this surface as a 2-torus, but they can be sorted out. Different values of g arise if we take different polygons, but every g occurs. In the jargon, a torus with two or more holes has a natural hyperbolic structure. So now we can reinterpret the list of standard surfaces:

- Sphere, $g = 0$: elliptic geometry.
- Torus, $g = 1$: Euclidean geometry.
- g-holed torus, $g = 2, 3, 4, \ldots$: hyperbolic geometry.

It may seem that we have thrown the baby out with the bathwater, because topology is supposed to be about rubber-sheet geometry, not rigid geometry. But now we can easily put the rubber back. Rigid geometry is used here only to *define* the standard surfaces. It provides simple descriptions, which happen to have extra rigid structure. Now relax the rigidity – in effect, allow the space to *become* like rubber. Let it deform in ways that rigidity prohibits. Now we get surfaces that are topologically equivalent to the standard ones, but are not equivalent by rigid motions. The classification theorem tells us that every topological surface can be obtained in this manner.

Topologists were aware of this link between geometry and the classification theorem for surfaces, but it looked like a funny coincidence, no doubt a consequence of the rather limited possibilities in two dimensions. Everyone knew that the 3-dimensional case was much richer, and in particular spaces of constant curvature did not exhaust the possibilities. It took one of the world's top geometers, William Thurston, to realise that rigid geometry might still be relevant to 3-dimensional topology. There were already a few hints: Poincaré's 3-sphere has a natural elliptic/ spherical geometry, coming from its definition. Although a standard dodecahedron lives in Euclidean space, the angle between adjacent faces is less than 120 degrees, so three such angles do not make a full circle. To remedy that, we have to inflate the dodecahedron so that its faces bulge slightly: this makes the natural geometry spherical, not Euclidean. Analogously, triangles on a sphere also bulge. The 3-torus, obtained by identifying opposite faces of a cube, has a flat – that is, Euclidean – geometry, just like its 2-dimensional analogue. Max Dehn and others had discovered a few 3-dimensional topological spaces with natural hyperbolic geometries.

Thurston began to see hints of a general theory, but two innovations were needed to make it even remotely plausible. First, the range of 3-dimensional geometries had to be extended. Thurston wrote down reasonable conditions, and proved that exactly eight geometries satisfy them. Three of them are the classics: spherical, Euclidean, and hyperbolic geometry. Two more are like cylinders: flat in one direction, curved in two other directions. The curved part is either positively curved, the 2-sphere, or negatively curved, the hyperbolic plane. Finally, there are three other, rather technical, geometries.

Second: some 3-dimensional spaces did not support any of the eight geometries. The answer was to cut the space into pieces. One piece might have a spherical geometric structure, another a hyperbolic one, and so on. In order to be useful, the cutting had to be done in a very tightly constrained manner, so that reassembling the pieces conveyed useful information. The good news was that in many examples, this turned out to be possible. In 1982, in a great leap of imagination, Thurston stated his geometrisation conjecture: *every* 3-dimensional space can be cut up, in an essentially unique manner, into pieces, each of which has a natural geometric structure corresponding to one of the eight possible geometries. He also proved that if his geometrisation conjecture were true, then the Poincaré conjecture would be a simple consequence.

Meanwhile, a second line of attack was emerging, also geometric, also based on curvature, but coming from a very different area: mathematical physics. Gauss, Riemann, and a school of Italian geometers had developed a general theory of curved spaces, called manifolds, with a concept of distance that extended Euclidean and classical non-Euclidean geometry enormously. Curvature need no longer be constant: it could vary smoothly from one point to another. A shape like a dog's bone, for instance, is positively curved at each end, but negatively curved in between, and the amount of curvature varies smoothly from one region to the next. Curvature is quantified using mathematical gadgets known as tensors. Around 1915 Albert Einstein realised that curvature tensors were exactly what he needed to extend his theory of special relativity, which was about space and time, to general relativity, which also included gravity. In this theory, the gravitational field is represented as the curvature of space, and

Einstein's field equations describe how the associated measure of curvature, the curvature tensor, changes in response to the distribution of matter. In effect, the curvature of space *flows* as time passes; the universe or some part of it spontaneously changes its shape.

Richard Hamilton, a specialist in Riemannian geometry, realised that the same trick might apply more generally, and that it might lead to a proof of the Poincaré conjecture. The idea was to work with one of the simplest measures of curvature, called Ricci curvature after the Italian geometer Gregorio Ricci-Curbastro. Hamilton wrote down an equation that specified how Ricci curvature should change over time: the Ricci flow. The equation was set up so that the curvature should gradually redistribute itself in as even a manner as possible. This is a bit like the cat under a carpet in chapter 4, but now, even though the cat can't escape, it can smear itself out in an even layer. (A topological cat is essential here.)

For example, in the 2-dimensional case, start with a pear-shaped surface, Figure 41. This has a region at one end that is strongly and positively curved; a region at the other, fatter, end that is also positively curved, but not so strongly; and a band in between where the curvature is negative. The Ricci flow in effect transports curvature from the strongly curved end (and to a lesser extent from the other end) into the negatively curved band, until all of the negative curvature has been gobbled up. At that stage the result is a bumpy surface with positive curvature everywhere. The Ricci flow continues to redistribute curvature, taking it away from the highly curved regions and moving it to the less curved regions. As time becomes ever larger, the surface gets closer and closer to one that has constant positive curvature – that is, a Euclidean sphere. The topology remains the same, even though the detailed shape changes, so we can prove that the original pear-shaped surface is topologically equivalent to a sphere by following the Ricci flow.

Fig 41 How the Ricci flow turns a pear into a sphere.

In this example the topological type of the surface was obvious to start with, but the same general strategy works for any manifold. Start with a complicated shape and follow the Ricci flow. As time passes, the curvature redistributes itself more evenly, and the shape becomes simpler. Ultimately, you should end up with the simplest shape having the same topology as the original manifold, whatever that may be. In 1981 Hamilton proved that this strategy works in 2 dimensions, providing a new proof of the classification theorem for surfaces.

He also made significant progress on the analogous strategy for 3-dimensional manifolds, but now there was a serious obstacle. In 2 dimensions, every surface automatically simplifies itself by following the Ricci flow. The same is true in 3 dimensions if the initial manifold has strictly positive curvature at every point: never zero or negative. Unfortunately, if there are points at which the curvature is zero, and there often are, the space can get tangled up in itself as it flows. This creates singularities: places where the manifold ceases to be smooth. At such points, the equation for the Ricci flow breaks down, and the redistribution of curvature has to stop. The natural way to get round this obstacle is to understand what the singularities look like and redesign the manifold – perhaps by cutting it into pieces – so that the Ricci flow can be given a jump-start. Provided you have enough control over how the topology of the remodelled manifold relates to that of the original one, this modified strategy can be successful. Unfortunately, Hamilton also realised that for 3-dimensional spaces, the singularities in the Ricci flow can be very complicated indeed – apparently too complicated to use that kind of trick. The Ricci flow quickly became a standard technique in geometry, but it fell short of proving the Poincaré conjecture.

By 2000 mathematicians had still not cracked the conjecture, and its importance was recognised more widely when it was made one of the seven millennium problems. By then it had also become clear that if Hamilton's idea could somehow be made to work in sufficient generality, it wouldn't just imply the Poincaré conjecture. It would prove Thurston's geometrisation conjecture as well. The prize was glittering, but it remained tantalisingly out of reach.

Mathematics is like the other branches of science: in order for research to be accepted as correct, it has to be published, and for that to happen, it has to survive peer review. Experts in the field have to read the paper carefully, check the logic, and make sure the calculations are correct. This process can take a long time for a complicated and important piece of mathematics. As mentioned in chapter 1, the remedy used to be a preprint, but nowadays there is a standard website, the arXiv ('archive'), where electronic preprints can be posted, subject to a partial refereeing process and an endorsement procedure to weed out rubbish. Today, most researchers first encounter new results on the arXiv or the author's own website.

In 2002 Grigori Perelman put a preprint on the arXiv about the Ricci flow. It made a remarkable claim: the flow is gradient-like. That is, there is a well-defined 'downhill' direction, a single numerical quantity associated with the shape of the manifold, and the manifold flows downhill in the sense that this quantity always decreases as time passes. It's analogous to height in a landscape, and it provides a quantitative measure of what 'simplifying' a manifold means. Gradient-like flows are fairly restricted: they can't go round and round in circles or behave chaotically. No one seems to have suspected that the Ricci flow would be so tame. But Perelman didn't just make the claim: he proved it. He ended by outlining an argument that would prove the Thurston geometrisation conjecture – which implies the Poincaré conjecture but goes much further – promising further details in subsequent postings to the arXiv. Over the next eight months he posted two follow-up papers containing many of the promised details.

The first posting caused quite a stir. Perelman was claiming to have carried out the entire Hamilton programme, by using the Ricci flow to simplify a 3-dimensional manifold and proving that the result was exactly what Thurston had predicted. The other two postings added further weight to the feeling that Perelman knew what he was talking about, and that his ideas went well beyond outlining a plausible strategy with the odd logical gap or unproved assumption. The usual scepticism of the mathematical community towards claims of a solution to a great problem was muted; there was a general feeling that he might well have succeeded.

However, the devil is in the detail, and in mathematics the detail can be devilish indeed. The work had to be checked, at length and in depth, by people who understood the areas involved and were aware of the potential pitfalls. And that wasn't straightforward, because Perelman had combined at least four very different areas of mathematics and mathematical physics, and few people understood more than one or two of them. Deciding whether his proof was correct would need a lot of teamwork and a lot of effort. Moreover, the preprints on the arXiv did not include full details at the normal level for a published paper. They were pretty clearly written, for preprints, but they didn't always dot the i's and cross the t's. So the experts had to reconstruct a certain amount of Perelman's thinking – and he had been deeply immersed in the work for years.

It all took time. Perelman lectured on his proof and answered e-mails questioning various steps. Whenever anyone found what seemed to be a gap, he responded quickly with further explanation, filling it. The signs were encouraging. But no one was going to risk their reputation by stating publicly that Perelman had proved the Poincaré conjecture, let alone the more difficult geometrisation conjecture, until they were very confident that there were no mistakes in the proof. So despite the generally favourable view of Perelman's work, public acceptance was initially withheld. This was unavoidable, but also unfortunate, because as the waiting dragged on, Perelman became increasingly irritated by what seemed to be fence-sitting. He *knew* his proof was correct. He understood it so well that he couldn't see why others were having trouble. He declined to write up the work in more detail or to submit it to a journal. As far as he was concerned, it was a done deal, and the arXiv preprints contained everything that was required. He stopped answering questions about allegedly missing details. To him, they weren't missing. Come on guys, you can figure this out without further help from me. It's not that difficult.

Some reports have suggested that in this respect the mathematical community was unfair to Perelman. But this misunderstands how the mathematical community functions when great problems are allegedly solved. It would have been irresponsible to just slap him on the back, say 'well done!', and ignore the missing steps in his preprints. It was entirely proper, indeed unavoidable, to ask him to prepare more extensive treatments, suitable for publication. On a problem of this importance, a rush job is dangerous and unacceptable. The experts went out of their way to spend a lot of time on Perelman's proof, and they kept their natural scepticism at bay to an unusual extent. His treatment was if anything *more* favourable than usual. And eventually, when this process was complete, his work was accepted as correct.

By that time, however, Perelman had lost patience. It perhaps didn't help that he had solved such an important problem that nothing else was likely to match it. He was like a mountaineer who has climbed Everest solo without oxygen. There were no comparable challenges left. Media publicity repelled him: he wanted acceptance by his peers, not by television presenters. So it is not terribly surprising that when his peers finally agreed he was right, and offered him a Fields medal and the Clay prize, he didn't want to know.

Perelman's proof is deep and elegant, and opens up a new world of topology. It implements Hamilton's Ricci flow programme by finding clever ways to get round the occurrence of singularities. One is to change the scales of space and time to get rid of the singularity. When this approach fails, the singularity is said to collapse. In such cases, he analyses the geometry of the Ricci flow in some detail, classifying how a collapse can occur. In effect, the space puts out ever-thinner tentacles, perhaps in profusion, like the branches of a tree. Whenever a tentacle is close enough to collapsing, it can be cut, and the spiky, sharply curved end can be cut off and replaced by a smooth cap. For some of these tentacles, the Ricci flow grinds to a halt: if so, leave them alone. If not, the Ricci flow can be restarted. So some tentacles end in smooth caps, and others are temporarily interrupted, but continue to flow.

This cut-and-paste capping procedure chops the space up in much the same way as Thurston's dissection into pieces, each with one of his eight geometries, and the two procedures turn out to give more or less identical results. One technical point is vital: the capping-off operations do not pile up ever faster, so that infinitely many of them occur in a finite time. This is one of the most complicated parts of the proof.

Some commentators have criticised the mathematical community for treating Perelman unfairly. No one should be immune to criticism, and there were a few incidents that might well be classed as unfair or in other ways ill-considered, but the mathematical community reacted rapidly and positively to Perelman's work. It also reacted cautiously, which is absolutely standard in mathematics and science, for excellent reasons. The inevitable glare of publicity, heightened by the million-dollar prize, had an impact on everyone, Perelman included.

From Perelman's first posting on the arXiv in November 2002 to the announcement in March 2010 that he had been awarded the Clay prize took eight years. That sounds like a lengthy delay, perhaps an unreasonable one. However, that first posting tackled only part of the problem. Most of the rest was posted in March 2003. By September 2004, eighteen months after this second posting, the Ricci flow and topological communities had already worked through the proof – a process that had started mere days after the first posting – and the main experts announced that they 'understood the proof'. They had found mistakes, they had found gaps, but they were convinced that those could all be put right. Eighteen months is actually remarkably quick when something so important is at stake.

Late in 2005 the International Mathematical Union approached Perelman and offered him a Fields medal, the subject's highest honour, to be awarded at the International Congress of Mathematicians in 2006. The ICM is held every four years, so this was the first opportunity to recognise his work in this manner. Because a few doubts remained about the complete proof of the Poincaré conjecture – mistakes were still turning up – the medal was officially awarded for advances in understanding the Ricci flow, the part of Perelman's preprints that were now considered error-free.

The conditions for the award of the prize are stated on the Clay Institute's website. In particular, a proposed solution has to be published in a refereed journal and still be accepted by the mathematical community two years later. After that, a special advisory committee looks into the matter and recommends whether or not to award the prize. Perelman has not complied with the first condition, and it doesn't seem likely that he ever will. In his view, the arXiv preprints suffice. Nevertheless, the Clay Institute waived that requirement and started the statutory two-year wait to see whether any mistakes or other issues emerged. That ended in 2008, after which the Institute's procedures, carefully structured to avoid awarding the prize prematurely, had to be followed.

It is true that some experts were slow to express their belief that the proof was correct. The reason is straightforward: they were genuinely uncertain. It's no great exaggeration to say that the only person capable of grasping Perelman's proof quickly was another Perelman. You can't read a mathematical proof like a musician sight-reads a score. You have to convince yourself it all makes sense. Whenever the argument gets very complicated, you know there is a serious chance of a mistake. The same applies when the ideas get too simple; many a prospective proof has fallen foul of an assertion so evident that no proof seemed necessary. Until the experts were genuinely sure that proof was basically correct – at which point they gave Perelman full credit *despite* the remaining gaps and errors – it was sensible to suspend judgement. Think about all the fuss over the eventually discredited work on cold fusion. Caution is the correct professional response, and the cliché applies: extraordinary claims require extraordinary evidence.

Why did Perelman reject the Fields medal and decline the Clay prize? He alone knows, but he wasn't interested in that kind of recognition and he repeatedly said so. He had already refused smaller prizes. He made it clear at the start that he did not want premature publicity; ironically, this is the same reason why the experts were understandably reluctant to take the plunge too soon. To be realistic, there wasn't the slightest chance that the media would *not* notice his work. For years the mathematical community has been making a big effort to get newspapers, radio, and television interested in the subject. It doesn't make a lot of sense to complain when this effort succeeds, or to expect the media to ignore the hottest mathematics story since Fermat's last theorem. But Perelman didn't see it that way, and he retreated into his shell. There is an offer on the table to make the prize money available for educational or other purposes, if he agrees. So far, he has failed to respond.

11

They can't all be easy
P/NP Problem

OWADAYS, MATHEMATICIANS ROUTINELY use computers to solve problems, even great problems. Computers are good at arithmetic, but mathematics goes far beyond mere 'sums', so putting a problem on a computer is seldom straightforward. Often the hardest part of the work is to convert the problem into one that a computer calculation can solve, and even then the computer may struggle. Many of the great problems that have been solved recently involve little or no work with a computer. Fermat's last theorem and the Poincaré conjecture are examples.

When computers have been used to solve great problems, like the four colour theorem or the Kepler conjecture, the computer effectively plays the role of servant. But sometimes the roles are reversed, with mathematics as the servant of computer science. Most of the early work on computer design made good use of mathematical insights, for example the connection between Boolean algebra – an algebraic formulation of logic – and switching circuits, developed in particular by the engineer Claude Shannon, the inventor of information theory. Today, both practical and theoretical aspects of computers rely on the extensive use of mathematics, from many different areas.

One of the Clay millennium problems lies in the borderland of mathematics and computer science. It can be viewed both ways: computer science as a servant of mathematics, and mathematics as a servant of computer science. What it requires, and is helping to bring about, is more balanced: a partnership. The problem is about computer algorithms, the mathematical skeletons from which computer programs are made. The crucial concept here is how efficient the algorithm is: how many computational steps it takes to get an answer for a given amount of input data. In practical terms, this tells us how long the computer will take to solve a problem of given size.

The word algorithm goes back to the Middle Ages, when Muhammad ibn Mūsā al-Khwārizmī wrote one of the earliest books on algebra. Earlier, Diophantus had introduced one element we associate with algebra: symbols. However, he used the symbols as abbreviations, and his methods for solving equations were presented through specific – though typical – examples. Where we would now write something like '$x + a = y$, therefore $x = y - a$', Diophantus would write 'suppose $x + 3 = 10$, then $x = 10 - 3 = 7$' and expect his readers to understand that the same idea would work if 3 and 10 were replaced by any other numbers. He would explain his illustrative example using symbols, but he wouldn't manipulate the symbols as such. Al-Khwārizmī made the general recipe explicit. He did this using words, not symbols, but he had the basic idea, and is generally considered to be the father of algebra. In fact, that name comes from the title of his book: *al-Kitāb al-Mukhtasar fī Hisāb al-Jabr wa'l-Muqābala* ('The compendious book on calculation by completion and balancing'). Al-jabr became algebra. The word 'algorithm' comes from a medieval version of his name, Algorismus, and it is now used to mean a specific mathematical process for solving a problem, one that is guaranteed to find the solution provided you wait long enough.

Traditionally, mathematicians considered a problem to be solved if, in principle, they could write down an algorithm leading to an answer. They seldom used that word, preferring to present, say, a formula for the solution, which is a particular kind of algorithm in symbolic language. Whether it was possible to apply the formula in practice wasn't terribly important: the formula *was* the solution. But the use of computers changed that view, because formulas that had been too complicated for hand calculation might become practical with the aid of a computer. However, it was a bit disappointing to find, as sometimes happened, that the formula was still too complicated: although the computer could try to run through the algorithm, it was too slow to reach the answer. So attention shifted to finding efficient algorithms. Both mathematicians and computer scientists had a vested interest in developing algorithms that really did give answers in a reasonable period of time.

Given an algorithm, it is relatively straightforward to estimate how long it will take (measured by the number of computational steps required) to solve a problem with a given size of input. This may require a certain amount of technique, but you know what process is involved and you know a lot about what it is doing. It is much more difficult to devise a more efficient algorithm if the one you start from turns out to be inefficient. And it is even harder to decide how good or bad the most efficient algorithm for a given problem can be, because that involves contemplating all possible algorithms, and you don't know what these are.

Early work on such questions led to a coarse but convenient dichotomy between algorithms that were efficient, in a simple but rough and ready sense, and those that were not. If the length of the computation grows relatively slowly as the size of the input increases, the algorithm is efficient and the problem is easy. If the length of the computation grows ever faster as the size of the input increases, the algorithm is inefficient and the problem is hard. Experience tells us that although some problems are easy, in this sense, most seem to be hard. Indeed, if all mathematical problems were easy, mathematicians would be out of a job. The millennium prize problem asks for a rigorous proof that at least one hard problem exists – or that contrary to experience, all problems are easy. It is known as the P/NP problem, and no one has a clue how to solve it.

We've already encountered a rough-and-ready measure of efficiency in chapter 2. An algorithm is class P if it has polynomial running time. In other words, the number of steps it takes to get the answer is proportional to some fixed power, such as the square or cube, of the size of the input data. Such algorithms are efficient, speaking very broadly. If the input is a number, that size is how many digits it has, not the number itself. The reason is that the quantity of information needed to specify the number is the space it occupies in the computer's memory, which is (proportional to) the number of digits. A problem is class P if there exists a class P algorithm that solves it.

Any other algorithm or problem belongs to the class not-P, and most of these are inefficient. Among them are those whose running time is exponential in the input data: approximately equal to some fixed number raised to the power of the size of the input. These are class E, and they are definitely inefficient.

Some algorithms are so efficient that they run much faster than polynomial time. For example, to determine whether a number is even or odd, look at its last digit. If (in decimal notation) this is 0, 2, 4, 6, or 8, the number is even; otherwise it is odd. The algorithm has at most six steps:

Is the last digit 0? If yes, then STOP. The number is even.
Is the last digit 2? If yes, then STOP. The number is even.
Is the last digit 4? If yes, then STOP. The number is even.
Is the last digit 6? If yes, then STOP. The number is even.
Is the last digit 8? If yes, then STOP. The number is even.
STOP. The number is odd.

So the running time is at most 6, independently of the input size. It belongs to the class 'constant time'.

Sorting a list of words into alphabetical order is a class P problem. A straightforward way to perform this task is the bubble sort, so named because in effect words move up the list like bubbles in a glass of fizzy drink if they are further down the list than words that should be below them in alphabetical order. The algorithm repeatedly works through the list, compares adjacent words, and swaps them if they are in the wrong order. For example, suppose the list starts out as

PIG DOG CAT APE

On the first run through this becomes

DOG PIG CAT APE
DOG **CAT PIG** APE
DOG CAT **APE PIG**

where the bold words are the ones that have just been compared. On the second run this becomes

CAT DOG APE PIG

CAT **APE DOG** PIG
CAT APE **DOG PIG**

The third run goes

APE CAT DOG PIG
APE **CAT DOG** PIG
APE CAT **DOG PIG**

On the fourth run, nothing moves so we know we've finished. Notice how APE bubbles up step by step to the top (that is, front).

With four words, the algorithm runs through three comparisons at each stage, and there are four stages. With n words, there are $n - 1$ comparisons per stage and n stages, a total of $n(n - 1)$ steps. This is a little less than n^2, so the running time is polynomial, indeed quadratic. The algorithm may terminate sooner, but in the worst case, when the words are in exactly the reverse order, it takes $n(n - 1)$ steps. The bubble sort is obvious and class P, but it is nowhere near the most efficient sorting algorithm. The fastest comparison sort, which is set up in a more clever way, runs in $n \log n$ steps.

A simple algorithm with exponential running time, class E, is 'print a list of all binary numbers with n digits.' There are 2^n numbers in the list, and printing each (and calculating it) takes roughly n steps, so the running time is roughly $2^n n$, which is bigger than 2^n but less than 3^n when n is sufficiently large. However, this example is a bit silly because what makes it so slow is the size of the output, not the complexity of the calculation, and this observation will turn out to be crucial later on.

A more typical class E algorithm solves the travelling salesman problem. A salesman has to visit a number of cities. He can do so in any order. Which route visits them all in the shortest total distance? The naive way to solve this is to list all possible routes, calculate the total distance for each, and find the shortest one. With n cities there are

$$n! = n \times (n - 1) \times (n - 2) \times \cdots \times 3 \times 2 \times 1$$

routes (read 'factorial n'). This grows faster than any exponential.[74] A more efficient method, called dynamic programming, solves the travelling salesman problem in exponential time. The first such method, the Held-Karp algorithm, finds the shortest tour in $2^n n^2$ steps, which again lies between 2^n and 3^n when n is sufficiently large.

Even though these algorithms are 'inefficient', special tricks can be used to shorten the computation when the number of cities is large by human standards, but not too large for the tricks to cease to be effective. In 2006 D.L. Applegate, R.M. Bixby, V. Chvátal, and W.J. Cook solved the travelling salesman problem for 85,900 cities, and this was still the record in mid-2012.[75]

These examples of algorithms don't just illustrate the concept of efficiency. They also drive home my point about the difficulty of finding improved algorithms, and the even greater difficulty of finding one that is as efficient as possible. All of the known algorithms for the travelling salesman problem are class E, exponential time – but that does not imply that no efficient algorithm exists. It just shows that we haven't found one yet. There are two possibilities: we haven't found a better algorithm because we're not clever enough, or we haven't found a better algorithm because no such thing exists.

Chapter 2 is a case in point. Until Agrawal's team found their class P algorithm for primality testing, the best known algorithm was not-P. It was still pretty good, with running time $n \log n$ for n-digit numbers, which is actually better than the Agrawal-Kayal-Saxena algorithm until we reach numbers with 10^{1000} digits. Before their algorithm was discovered, opinion on the status of primality testing was divided. Some experts suspected it was class P and a suitable algorithm would be found; some thought it wasn't. The new algorithm came from nowhere, one of thousands of ideas that someone could have tried; this one happened to work. The precedent here is sobering: we don't know, we can't tell, and the experts' best guess may or may not be a good one.

The great problem that concerns us here asks for the answer to a more fundamental question. Are there any hard problems? Might all problems be easy, if only we were clever enough? The actual statement is subtler, because we've already seen one instance of a problem that is indubitably hard: print a list of all binary numbers with n digits. As I remarked, this is a bit silly: the difficulty lies not in the calculation, but the sheer drudgery of printing out a very long answer. We know there's no short cut, because the answer is that long *by definition*. If it were shorter, it wouldn't be the answer.

In order to pose a sensible question, trivial examples like this one have to be eliminated. The way to do that is to introduce another class of algorithm, class NP. This is not the class not-P; it is the class of algorithms that run in nondeterministic polynomial time. The jargon means that however long the algorithm takes to come up with its answer, we can *check that the answer is right* in polynomial time. Finding the answer may be hard, but once found, there's an easy test of its validity.

The word 'nondeterministic' is used here because it's possible to solve an NP problem by making an inspired guess. Having done so, you can confirm that it really is correct (or not). For instance, if the problem is to factorise the number 11,111,111,111, you might guess that one factor is the prime number 21,649. As it stands, that's just a wild guess. But it's easy to check: just divide by it and see what you get. The result turns out to be 513,239, exactly, no remainder. So the guess was correct. If I'd guessed 21,647 – which is also prime – instead, then division would lead to the result 513,286 plus a remainder of 9069. So that guess would have been wrong.

Making a correct guess is basically a miracle here, or else there's a trick (I worked out the factors of 11,111,111,111 before 'guessing'). But that's actually what we want. If it weren't miraculous, you could turn a class NP algorithm into a class P algorithm by just making lots and lots of guesses until one turns out to be right. My example suggests why this won't work: you need too many guesses. Indeed, all we are doing here is 'trial division' by all possible primes, until one works. We know from chapter 2 that this is a hopeless way to find factors.

Class NP rules out silly examples like my very long list. If someone guesses a list of all binary digits of length n, then it doesn't just take exponential time to print the list out. It also takes exponential time to read it, so it takes even longer to check it's correct. It would be a truly horrible proofreading task. Class P is definitely contained in class NP. If you can find the answer in polynomial time, with a guarantee that it's correct, then you've already checked it. So the check automatically requires nothing worse than polynomial time. If someone presented you with the supposed answer, you could just run the entire algorithm again. That's the check.

Now we can state the millennium problem. Is NP bigger than P, or are they the same? More briefly: is P equal to NP?

If the answer is 'yes' then it would be possible to find fast, efficient algorithms for scheduling airline flights, optimising factory output, or performing a million other important practical tasks. If the answer is 'no', we will have a cast-iron guarantee that all of the apparently hard problems really are hard, so we will be able to stop wasting time trying to find fast algorithms for them. We win either way. What's a nuisance is not knowing which way it goes.

It would make life much simpler for mathematicians if the answer were 'yes', so the pessimist in every human being immediately suspects that life isn't going to be that simple, and the likely answer is 'no'. Otherwise we're all getting a free lunch, which we don't deserve and haven't earned. I suspect most mathematicians would actually prefer the answer to be 'no', because that should keep them in business until the end of civilisation. Mathematicians prove themselves by solving hard problems. For whatever reason, most mathematicians and computer scientists expect the answer to the question 'does P equal NP?' to be 'no'. Hardly anyone expects it to be 'yes'.

There are two other possibilities. It might be possible to prove that P is equal to NP without actually finding a polynomial-time algorithm for any specific NP problem. Mathematics has a habit of providing existence proofs that are not constructive; they show that something exists but don't tell us what it is. Examples include primality tests, which cheerfully inform us that a number is not prime without producing any specific factor, or theorems in number theory asserting that the solutions to some Diophantine equation are bounded – less than some limit – without providing any specific bound. A polynomial-time algorithm might be so complicated that it's impossible to write it down. Then the natural pessimism about free lunches would be justified, even if the answer turned out to be affirmative.

More drastically, some researchers speculate that the question may be undecidable within the current formal logical framework for mathematics. If so, neither 'yes' nor 'no' can be proved. Not because we're too stupid to find the proof: because there isn't one. This option became apparent in 1931 when Kurt Gödel set the cat of undecidability loose among the philosophical pigeons that infested the foundations of mathematics, by proving that some statements in arithmetic are undecidable. In 1936 Alan Turing found a simpler undecidable problem, the halting problem for Turing machines. Given an algorithm, is there always a proof that it either stops, or a proof that it must go on for ever? Turing's surprising answer was 'no'. For some algorithms no proof either way exists. The P/NP problem could, perhaps, be like that. It would explain why no one can either prove it or disprove it. But no one can prove or disprove that the P/NP problem is undecidable, either. Maybe its undecidability is undecidable ...

The most direct way to approach the P/NP problem would be to select some question known to be in class NP, assume that there exists a polynomial-time algorithm to solve it, and somehow derive a contradiction. For a time, people tried this technique on a variety of problems, but in 1971 Stephen Cook realised that the choice of problem often makes no difference. There is a sense in which all such problems – give or take some technicalities – stand or fall together. Cook introduced the notion of an NP-complete problem. This is a specific NP problem, with the property that if there exists a class P algorithm to solve it, then *any* NP problem can be solved using a class P algorithm.

Cook found several NP-complete problems, including SAT, the Boolean satisfiability problem. This asks whether a given logical expression can be made true by choosing the truth or falsity of its variables in a suitable manner. He also obtained a deeper result: a more restrictive problem, 3-SAT, is also NP-complete. Here the logical formula is one that can be written in the form 'A or B or C or ... or Z', where each of A, B, C, ..., Z is a logical formula involving only three variables. Not necessarily the same three variables each time, I hasten to add. Most proofs that a given problem is NP-complete trace back to Cook's theorem about 3-SAT.

Cook's definition implies that all NP-complete problems are on the same footing. Proving that one of them is class P would prove that all of them are class P. This result leaves open a tactical possibility: some NP-complete problems might be easier to work with than others. But strategically, it suggests that you may as well pick one specific NP-complete problem and work with that one. NP-complete problems all stand or fall together because an NP-complete problem can simulate any NP problem whatsoever. Any NP problem can be converted into a special case of the NP-complete one by 'encoding' it, using a code that can be implemented in polynomial time.

For a flavour of this procedure, consider a typical NP-complete problem: find a Hamiltonian cycle in a network. That is, specify a closed path along the edges of the network that visits every vertex (dot) exactly once. Closed means that the path returns to its starting point. The size of input data here is the number of edges, which is less than or equal to the square of the number of dots since each edge joins two dots. (We assume at most one edge joins a given pair.) No class P algorithm to solve this problem is known, but suppose, hypothetically, that there were one. Now choose some other problem, and call it problem X. Suppose that problem X can be rephrased in terms of finding such a path in some network associated to problem X. If the method for translating the data of problem X into data about the network, and conversely, can be carried out in polynomial time, then we automatically obtain a class P algorithm for problem X, like this:

1   Translate problem X into the search for a Hamiltonian cycle on the related network, which can be done in polynomial time.
2   Find such a cycle in polynomial time using the hypothetical algorithm for the network problem.
3   Translate the resulting Hamiltonian cycle back into a solution of problem X, which again can be done in polynomial time.

Since three polynomial-time steps combined run in polynomial time, this algorithm is class P.

To show how this works, I'll consider a less ambitious version of the Hamiltonian cycle problem in which the path is not required to be closed. This is called the Hamiltonian path problem. A network may possess a Hamiltonian path without possessing a cycle: Figure 42 (left) is an example. So a solution to the Hamiltonian cycle problem may not solve the Hamiltonian path problem. However, we can convert the Hamiltonian path problem into a Hamiltonian cycle problem on a related, but different, network. This is obtained by adding one extra dot, joined to every dot in the original network as in Figure 42 (right). Any Hamiltonian cycle in the new network can be converted to a

Hamiltonian path in the original one: just omit the new vertex and the two edges of the cycle that meet it. Conversely, any Hamiltonian path in the original network yields a Hamiltonian cycle in the new network: just join the two ends of the Hamiltonian path to the new dot. This 'encoding' of the path problem as a cycle problem introduces only one new dot and one new edge per dot in the original. So this procedure, and its inverse, run in polynomial time.

Fig 42 *Left*: Network with a Hamiltonian path (solid line) but no Hamiltonian cycle. *Right*: Add an extra dot (grey) and four more lines to convert the Hamiltonian path into a Hamiltonian cycle (solid line). The two grey edges are not in the cycle but are needed for the construction of the larger network.

Of course all I've done here is to encode one specific problem as a Hamiltonian cycle problem. To prove that the Hamiltonian cycle problem is NP-complete, we have to do the same for any NP problem. This can be done: the first proof was found by Richard Karp in 1972, in a famous paper that proved 21 different problems are NP-complete.[76]

The travelling salesman problem is 'almost' NP-complete, but there is a technical issue: it is not known to be NP. Over 300 specific NP-complete problems are known, in areas of mathematics that include logic, networks, combinatorics, and optimisation. Proving that any one of them can or cannot be solved in polynomial time would prove the same for every one of them. Despite this embarrassment of riches, the P/NP problem remains wide open. It wouldn't surprise me if it still is, a hundred years from now.

12

Fluid thinking
Navier-Stokes Equation

FIVE OF THE MILLENNIUM PROBLEMS, including the three discussed so far, come from pure mathematics, although the P/NP problem is also fundamental to computer science. The other two come from classical applied mathematics and modern mathematical physics. The applied mathematics problem arises from a standard equation for fluid flow, the Navier-Stokes equation, named after the French engineer and physicist Claude-Louis Navier and the Irish mathematician and physicist George Stokes. Their equation is a partial differential equation, which means that it involves the rate of change of the flow pattern in both space and time. Most of the great equations of classical applied mathematics and physics are also partial differential equations – we just met one, Laplace's – and those that are not are ordinary differential equations, involving only the rate of change with respect to time.

In chapter 8 we saw how the motion of the solar system is determined by Newton's laws of gravity and motion. These relate the accelerations of the Sun, Moon, and planets to the gravitational forces that are acting. Acceleration is the rate of change of velocity with respect to time, and velocity is the rate of change of position with respect to time. So this is an ordinary differential equation. As we saw, solving such equations can be very difficult. Solving partial differential equations is generally a lot harder.

For practical purposes, the equations for the solar system can be solved numerically using computers. That's still hard, but good methods now exist. The same is true for practical applications of the Navier-Stokes equations. The techniques employed are known as computational fluid dynamics, and they have a vast range of important applications: aircraft design, car aerodynamics, even medical problems like the flow of blood in the human body.

The millennium prize problem does not ask mathematicians to find explicit solutions to the Navier-Stokes equation, because this is essentially impossible. Neither is it about numerical methods for solving the equations, important though these are. Instead, it asks for a proof of a basic theoretical property: the *existence* of solutions. Given the state of a fluid at some instant of time – the pattern in which it is moving – does there exist a solution of the Navier-Stokes equation, valid for all future time, starting from the state concerned? Physical intuition suggests that the answer must surely be 'yes', because the equation is a very accurate model of the physics of real fluids. However, the mathematical issue of existence is not so clear-cut, and this basic property of the equation has never been proved. It might not even be true.

The Navier-Stokes equation describes how the pattern of fluid velocities changes with time, in given circumstances. The equation is often referred to using the plural, Navier-Stokes equations, but it's the same either way. The plural reflects the classical view: in three-dimensional space, velocity has three components, and classically each component contributed one equation, making three in all. In the modern view, there is one equation for the velocity *vector* (a quantity with both size and direction), but this equation can be applied to each of three components of the velocity. The Clay Institute website uses the classical terminology, but here I will follow the modern practice. I mention this to avoid possible confusion.

The equation dates from 1822, when Navier wrote down a partial differential equation for the flow of a viscous – sticky – fluid. Stokes's contributions occurred in 1842 and 1843. Euler had written down a partial differential equation for a fluid with zero viscosity – no stickinesss – in 1757. Although this equation remains useful, most real fluids including water and air are viscous, so Navier and Stokes modified the Euler equation to take account of viscosity. Both scientists derived essentially the same equation independently, so it is named after them both. Navier made some mathematical errors but ended up with the right answer; Stokes got the mathematics right, which is how we know Navier's answer is correct despite his mistake. In its most general form, the equation applies to compressible fluids like air. However, there is an important special case in which the fluid is assumed to be incompressible. Such a model applies to fluids like water, which do compress under huge forces, but only very slightly.

There are two ways to set up the mathematical description of fluid flow: you can either describe the path that each particle of fluid takes as time passes, or you can describe the velocity of the flow at each point in space and each instant of time. The two descriptions are related: given one, you can – with effort – deduce the other. Euler, Navier, and Stokes all used the second point of view, because it leads to an equation that is much more tractable mathematically. So their equations refer to the fluid's velocity field. At each fixed instant of time, the velocity field specifies the speed and direction of every particle of fluid. As time varies, this description may change. This is why rates of change both in space and in time occur in the equation.

The Navier-Stokes equation has an excellent physical pedigree. It is based on Newton's laws of motion, applied to each tiny particle (small region) of fluid, and expresses, in that context, the law of conservation of momentum. Each particle moves because forces act on it, and Newton's law of motion states that the particle's acceleration is proportional to the force. The main forces are friction, caused by viscosity, and pressure. There are also forces generated by the particle's acceleration. The equation follows classical practice, and treats the fluid as an infinitely divisible continuum. In particular it ignores the fluid's discrete atomic structure at very small scales.

Equations of themselves are of little value: you have to be able to solve them. For the Navier-Stokes equation, this means calculating the velocity field: the speed and direction of the fluid at each point in space and each instant of time. The equation provides constraints on these quantities, but it doesn't prescribe them directly. Instead, we have to apply the equation to relate future velocities to current ones. Partial differential equations like Navier-Stokes have many different solutions; indeed, infinitely many. This is no surprise: fluids can flow in many different ways; the flow over the surface of a car differs from that over the wings of an aircraft. There are two main ways to select a particular flow from this multitude of possibilities: initial conditions and boundary conditions.

Initial conditions specify the velocity field at some particular reference time, usually taken to be time zero. The physical idea is that once you know the velocity field at this instant, the Navier-Stokes equation determines the field a very short time later, in a unique way. If you start by giving the fluid a push, it keeps going while obeying the laws of physics. Boundary conditions are more useful in most applications, because it is difficult to set up initial conditions in a real fluid, and in any case these are not entirely appropriate to applications like car design. What matters there is the shape of the car. Viscous fluid sticks to surfaces. Mathematically, this feature is modelled by specifying the velocity on these surfaces, which form the boundary of the region occupied by the fluid, which is where the equation is valid. For example, we might require the velocity to be zero on the boundary, or whichever other condition best models reality.

Even when initial or boundary conditions are specified, it is highly unusual to be able to write down an explicit formula for the velocity field, because the Navier-Stokes equation is nonlinear. The sum of two solutions is not normally a solution. This is one reason why the three-body problem of chapter 8 is so hard – though not the sole reason, because the two-body problem is also nonlinear yet has an explicit solution.

For practical purposes, we can solve the Navier-Stokes equation on a computer, representing the velocity field as a list of numbers. This list can be turned into elegant graphics, and used to calculate quantities of interest to engineers, such as the stresses on aircraft wings. Since computers can't handle infinite lists of numbers, and they can't handle numbers to infinite accuracy, we have to replace the actual flow by a discrete approximation, that is, a list of numbers that samples the flow at finitely many locations and times. The big issue is to ensure that the approximation is good enough.

The usual approach is to divide space into a large number of small regions to form a computational grid. The velocity is calculated only for the points at the corners of the grid. The grid might be just an array of squares (or cubes in three dimensions), like a chessboard, but for cars and aircraft it has to be more complicated, with smaller regions near the boundary, to capture finer detail of the flow. The grid may be dynamic, changing shape as time passes. Time is generally assumed to progress in steps, which may all be the same size, or may change size according to the prevailing state of the calculation.

The basis of most numerical methods is the way 'rate of change' is defined in calculus. Suppose that an object moves from one location to another in a very short period of time. Then the rate of change of position – the velocity – is the change in position divided by the time taken, with a small error that vanishes as the time period gets smaller. So we can approximate the rate of change, which is what enters into the Navier-Stokes equation, by this ratio of the spatial change to the temporal change. In effect, the equation now tells us how to push a known initial state – a specified list of velocities – one time step into the future. Then we have to repeat the calculation many times, to see what happens further into the future. There is a similar way to approximate solutions when the one we want is determined by boundary conditions. There are also many sophisticated ways to achieve the same result more accurately.

The more finely the computational grid is divided, and the shorter the time intervals are, the more accurate the approximation becomes. However, the computation also takes longer. So there is a compromise between accuracy and speed. Broadly speaking, an approximate answer obtained by computer is likely to be acceptable provided the flow does not have significant features that are smaller than the grid size. There are two main types of fluid flow, laminar and turbulent. In laminar flow, the pattern of movement is smooth, and layers of fluid glide neatly past each other. Here a small enough grid ought to be suitable. Turbulent flow is more violent and frothy, and the fluid gets mixed up in extremely complex ways. In such circumstances, a discrete grid, however fine, could easily cause trouble.

One of the characteristics of turbulence is the occurrence of vortexes, like small whirlpools, and these can be very tiny indeed. A standard image of turbulence consists of a cascade of ever-smaller vortexes. Most of the fine detail is smaller than any practical grid. To get round this difficulty, engineers often resort to statistical models in questions about turbulent flow. Another worry is that the physical model of a continuum might be inappropriate for turbulent flow, because the vortexes may shrink to atomic sizes. However, comparisons between numerical calculations and experiments show that the Navier-Stokes equation is a very realistic and accurate model – so good that many engineering applications now rely solely on computational fluid dynamics, which is cheap, instead of performing experiments with scale models in wind tunnels, which are expensive. However, experimental checks like these are still used when human safety is vital, for example when designing aircraft.

In fact, the Navier-Stokes equation is so accurate that it even seems to apply when physics suggests that it should stand a reasonable chance of failing: turbulent flow. At least, that is the case if it can be solved accurately enough. The main problem is a practical one: numerical methods for solving the equation take enormous quantities of computer time when the flow becomes turbulent. And they always miss some small-scale structure.

Mathematicians are always uneasy when the main information they have about a problem is based on some kind of approximation. The millennium prize for the Navier-Stokes equation tackles one of the key theoretical issues. Its solution would reinforce the gut feeling that the numerical methods usually work very well. There is a subtle distinction between the approximations used by the computer, which make small changes to the equation, and the accuracy of the answer, which is about small changes to the solution. Is an exact answer to an approximate question the same as an approximate answer to the exact question? Sometimes the answer is 'no'. The exact flow for a fluid with very small viscosity, for instance, often differs from an approximate flow for a fluid with zero viscosity.

One step towards understanding these issues is so simple that it can easily be overlooked: proving that an exact solution exists. There has to be something for the computer calculations to be approximations *to*. This observation motivates the millennium prize for the Navier-Stokes equation. Its official description on the Clay Institute website consists of four problems. Solving any one of them is enough to win the prize. In all four, the fluid is assumed incompressible. They are:

1   *Existence and smoothness of solutions in three dimensions.* Here the fluid is assumed to fill the whole of infinite space. Given any

initial smooth velocity field, prove that a smooth solution to the equation exists for all positive times, coinciding with the specified initial field.

2   *Existence and smoothness of solutions in the three-dimensional flat torus.* The same question, but now assuming that space is a flat torus – a rectangular box with opposite faces identified. This version avoids potential problems caused by the infinite domain assumed in the first version, which does not match reality and might cause bad behaviour for silly reasons.

3   *Breakdown of solutions in three dimensions.* Prove that (1) is wrong. That is, find an initial field for which a smooth solution does not exist for all positive times, and prove that statement.

4   *Breakdown of solutions in the three-dimensional flat torus.* Prove that (2) is wrong.

The same problems remain open for the Euler equation, which is the same as the Navier-Stokes equation but assumes no viscosity, but no prize is on offer for the Euler equation.

The big difficulty here is that the flow under consideration is three-dimensional. There is an analogous equation for fluid flowing in a plane. Physically, this represents either a thin layer of fluid between two flat plates, assumed not to cause friction, or a pattern of flow in three dimensions in which the fluid moves in exactly the same manner along a system of parallel planes. In 1969 the Russian mathematician Olga Alexandrovna Ladyzhenskaya proved that (1) and (2) are true, while (3) and (4) are false, for the two-dimensional Navier-Stokes equation and the two-dimensional Euler equation.

Perhaps surprisingly, the proof is harder for the Euler equation, even though that equation is simpler than the Navier-Stokes equation, omitting terms involving viscosity. The reason is instructive. Viscosity 'damps down' bad behaviour in the solution, which potentially might lead to some kind of singularity that prevents the solution existing for all time. If the viscosity term is missing, no such damping occurs, and this shows up as mathematical issues in the existence proof.

Ladyzhenskaya made other vital contributions to our understanding of the Navier-Stokes equation, proving not only that solutions do exist but also that certain computational fluid dynamics schemes approximate them as accurately as we wish.

The millennium prize problems refer to incompressible flow because compressible flows are well known to be badly behaved. The equations for an aircraft, for example, run into all sorts of trouble if the aircraft travels faster than sound. This is the famous 'sound barrier', which worried engineers trying to design supersonic jet fighters, and the problem is related to the compressibility of air. If a body moves through an incompressible fluid, it pushes the fluid particles out of the way, like tunnelling through a box filled with ball bearings. If the particles pile up, they slow the body down. But in a compressible fluid, where there is a limit to the speed with which waves can travel – the speed of sound – that doesn't happen. At supersonic speeds, instead of being pushed out of the way, the air piles up ahead of the aircraft, and its density there increases without limit. The result is a shockwave. Mathematically, this is a discontinuity in the pressure of the air, which suddenly jumps from one value to a different one across the shockwave. Physically, the result is a sonic boom: a loud bang. If not understood and taken into account, a shockwave can damage the aircraft, so the engineers were right to worry. But the speed of sound isn't really a barrier, just an obstacle. The presence of shockwaves implies that the compressible Navier-Stokes equations need not have smooth solutions for all time, even in two dimensions. So the answer is already known in that case, and it's negative.

The mathematics of shockwaves is a substantial area within partial differential equations, despite this breakdown of solutions. Although the Navier-Stokes equation alone is not a good physical model for compressible fluids, it is possible to modify the mathematical model by adding extra conditions to the equations, which take discontinuities of shockwaves into account. But shockwaves do not occur in the flow of an incompressible fluid, so it is at least conceivable that in that context solutions should exist for all time, no matter how complicated the initial flow might be, provided it is smooth.

Some positive results are known for the three-dimensional Navier-Stokes equation. If the initial flow pattern involves sufficiently small velocities, so that the flow is very sluggish, then (1) and (2) are both true. Even if the velocities are large, (1) and (2) are true for some nonzero interval of time. There may not exist a solution valid for all future times, but there is a definite amount of time over which a solution does exist. It might seem that we can repeat this process, pushing a solution forward in time by small amounts, and then using the end result as a new initial condition. The problem with this line of reasoning is that the time intervals may shrink so rapidly that infinitely many such steps take a finite time. For instance, if each successive step takes half the time of the previous one, and the first step takes, say, 1 minute, then the whole process is over in a time $1 + \frac{1}{2} + \frac{1}{4} + \frac{1}{8} + \cdots$, which equals 2. If the solution ceases to exist – at the present time a purely hypothetical assumption, but one we can still contemplate – then the solution concerned is said to *blow up*. The time it takes for this to happen is the blowup time.

So the four questions ask whether solutions can, in fact, blow up. If they can't, (1) and (2) are true; if they can, (3) and (4) are. Perhaps solutions can blow up in an infinite domain, but not in a finite one. By the way, if the answer to (1) is 'yes' then so is the answer to (2), because we can interpret any flow pattern in a flat torus as a spatially periodic flow pattern in the whole of infinite space. The idea is to fill space with copies of the rectangular box involved, and copy the same flow pattern in each. The gluing rules for a torus ensure that the flow remains smooth when it crosses these flat interfaces. Similarly if the answer to (4) is 'yes' then so is the answer to (3), for the same reason. We just make the initial state spatially periodic. But for all we currently know, the answer to (2) might be 'yes' but the answer to (1) could be 'no'.

We do know one striking fact about blowups. If there is a solution with a finite blowup time, then the maximum velocity of the fluid, at all points in space, must become arbitrarily large. This could occur, for instance, if a jet of fluid forms, and the speed of the jet increases so rapidly that it diverges to infinity after a finite amount of time has passed.

These objections are not purely hypothetical. There are precedents for this kind of singular behaviour in other equations of classical mathematical physics. A remarkable example occurs in celestial mechanics. In 1988 Zhihong Xia proved that there exists an initial configuration of five point masses in three-dimensional space, obeying Newton's law of gravity, for which four particles disappear to infinity after a finite period of time – a form of blowup – and the fifth undergoes ever wilder oscillations. Earlier, Joseph Gerver had indicated that five bodies in a plane might all disappear to infinity in finite time, but he was unable to complete the proof for the scenario he envisaged. In 1989 he proved that this kind of escape definitely can occur in a plane if the number of bodies is large enough.

It is remarkable that this behaviour is possible, given that such systems obey the law of conservation of energy. Surely, if all bodies are moving arbitrarily fast, the total kinetic energy must increase? The answer is that there is also a decrease in potential energy, and for a point particle, the total gravitational potential energy is infinite. The bodies must also conserve angular momentum, but they can do that provided some of them move faster and faster in ever-decreasing circles.

The physical point involved is the famous slingshot effect, which is used routinely to dispatch space probes to distant worlds in the solar system. A good example is NASA's Galileo probe, whose mission was to travel to Jupiter, to study that giant planet and its many satellites. It launched in 1989 and arrived at Jupiter in 1995. One of the reasons it took so long was that its route was distinctly indirect. Although Jupiter's orbit lies outside that of the Earth, Galileo began by heading inwards, towards Venus. It passed close to Venus, returned to fly past the Earth, and headed out into space to look at the asteroid 951 Gaspra. Then it came back towards Earth, passed round our home planet *again*, and finally headed off towards Jupiter. Along the way it approached another asteroid, Ida, discovering that it had its own tiny moon, a new asteroid named Dactyl.

Why such a convoluted trajectory? Galileo gained energy, hence speed, from each close encounter. Imagine a space probe heading towards an approaching planet, not on a collision course, but getting pretty close to the surface, swinging round the back of the planet, and being flung off into space. As the probe passes behind the planet, each attracts the other. In fact, they've been attracting each other all along, but at this stage the force of attraction is at its greatest and so has the greatest effect. The planet's gravity gives the probe a speed boost. Energy must be conserved, so in compensation the probe slows the planet down very slightly in its orbit round the Sun. Since the probe has a very small mass and the planet has a very large mass, the effect on the planet is negligible. The effect on the probe is not: it can speed up dramatically.

Galileo got within 16,000 kilometres of Venus's surface, and gained 2.23 kilometres per second in speed. It then passed within 960 kilometres of Earth, and again within 300 kilometres, adding a further 3.7 kilometres per second. These manoeuvres were essential to get it to Jupiter, because its rockets were not powerful enough to take it there directly. The original plan had been to do just that, using the Centaur-G liquid-hydrogen-fuelled booster. But the disaster in which the space shuttle *Challenger* exploded shortly after takeoff caused this plan to be abandoned, because the Centaur-G was prohibited. So Galileo had to use a weaker solid-fuel booster instead. The mission was a huge success, and the scientific payoff included observing the collision between comet Shoemaker-Levy 9 and Jupiter in 1994, while the probe was still *en route* to Jupiter.

Xia's scenario also makes use of the slingshot effect. Four planets of equal mass form two close pairs, revolving round their common centres of mass in two parallel planes.[77] These two-body racquets play celestial tennis with a fifth, lighter body that shuttles to and fro between them at right angles to those planes. The system is set up so that every time this 'tennis ball' passes by a pair of planets, slingshot effects speed up the ball, and push the pair of planets outwards along the line joining the two pairs, so the tennis court gets longer and the players get further apart. Energy and momentum are kept in balance because the two planets concerned get slightly closer together, and revolve ever faster around their centre of mass. With the right initial setup, the pairs of planets move apart ever faster, and their speed increases so rapidly that they get to infinity after a finite amount of time. Meanwhile the tennis ball oscillates between them ever faster. Gerver's escape scenarios also use the slingshot effect.

Is this disappearing act relevant to real celestial bodies? Not if taken literally. It relies on the bodies being point masses. For many problems in celestial mechanics, that's a sensible approximation, but not if the bodies get arbitrarily close to each other. If bodies of finite size did that, they would eventually collide. Relativistic effects would stop the bodies moving faster than light, and change the law of gravity.

Anyway, the initial conditions, and assumptions that some masses are identical, would be too rare to happen in practice. Nonetheless, these curious examples show that even though the equations of celestial mechanics model reality very well in most circumstances, they can have complicated singularities that prevent solutions existing for all time. It has also recently been realised that slingshot effects in triple-star systems, where three stars orbit each other in complicated paths, can expel one of the stars at high velocity. So innumerable orphan stars, flung out of their systems by their siblings, may be roaming the galaxy – or even intergalactic space – cold, lonely, unwanted, and unnoticed.

When a differential equation behaves so strangely that its solutions cease to make sense after some finite period of time, we say that there is a singularity. The above work on the many-body problem is really about various types of singularity. The millennium prize problem about the Navier-Stokes equation asks whether singularities can occur in initial-value problems, for a fluid occupying either the whole of space or a flat torus. If a singularity can form in finite time, the result is likely to be blowup, unless the singularity somehow unravels itself subsequently, which seems unlikely.

There are two main ways to approach these questions. We can try to prove that singularities never occur, or we can try to find one by choosing suitable initial conditions. Numerical solutions can help, either way: they can suggest useful general features of flows, and they can provide strong hints about the possible nature of potential singularities. However, the potential lack of accuracy in numerical solutions means that any such hints must be treated with caution and justified more rigorously.

Attempts to prove regularity – the absence of singularities – employ a variety of methods to gain control over the flow. These include complicated estimates of how big or small certain key variables can become, or more abstract techniques. A popular approach is by way of so-called weak solutions, which are not exactly flows at all, but more general mathematical structures with some of the properties of flows. It is known, for instance, that the set of singularities of a weak solution of the three-dimensional Navier-Stokes equations is always small, in a specific technical sense.

Many different scenarios that might lead to singularities have been investigated. The standard model of turbulence as a cascade of ever-decreasing vortexes goes back to Andrei Kolmogorov in 1941, and he suggested that on very small scales, all forms of turbulence look very similar. The proportions of vortexes of given size, for example, follow a universal law. It is now known that as the vortexes become smaller, they change shape, and get longer and thinner, forming filaments. The law of conservation of angular momentum implies that the vorticity – how much the vortex is spinning – must increase. This is called vortex-stretching, and it is the kind of behaviour that might cause a singularity – for example, if the very small vortexes could become infinitely long in finite time, and the vorticity could become infinite at some points.

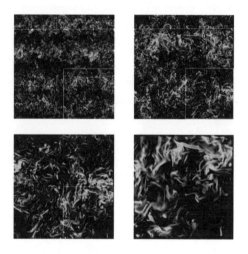

Fig 43 Zooms into a turbulent flow, simulated with the computer system VAPOR.

Figure 43 shows a zoom into very small scales of a turbulent flow, simulated by Pablo Mininni and colleagues using VAPOR, the Visualization and Analysis Platform for Ocean, Atmosphere, and Solar Research. The images show the vorticity intensity: how rapidly the fluid is spinning. They illustrate the formation of vortex filaments, the long thin structures in the figures, and show that they can cluster together to form larger-scale patterns. Their setup can perform simulations on cubic grids with more than 3 billion grid points.

In his article about this problem on the Clay Institute website,[78] Charles Fefferman writes:

> There are many fascinating problems and conjectures about the behavior of solutions of the Euler and Navier–Stokes equations... Since we don't even know whether these solutions exist, our understanding is at a very primitive level. Standard methods from [partial differential equations] appear inadequate to settle the problem. Instead, we probably need some deep, new ideas.

The complexity of the flow in images like Figure 43 drives home the difficulties that are likely to be encountered when seeking these ideas. Undaunted, mathematicians are soldiering on, seeking simple principles within the apparent complexities.

Quantum conundrum
Mass Gap Hypothesis

A FEW KILOMETRES NORTH OF GENEVA there is a sharp kink in the border between Switzerland and France. On the surface, all you see are minor roads and small villages. But between 50 and 175 metres underground is the largest scientific instrument on the planet. It is a gigantic circular tunnel, over 8 kilometres in diameter, joined to a second circular tunnel about a quarter as big. Most of it is under France but two sections are in Switzerland. Within the tunnels run pairs of pipes, which meet at four points.

It is the Large Hadron Collider, it cost €7.5 billion (about $9 billion) and it is probing the frontiers of particle physics. The key aim of the 10,000 scientists from over 100 nations who collaborated on it was to find the Higgs boson – or not to find it, if that's the way the continuum crumbled. They were looking for it to complete the Standard Model of particle physics, in which everything in the universe is made from 17 different fundamental particles. According to theory, the Higgs boson is what gives all particles mass.

In December 2011 ATLAS and CMS, two experimental divisions of the Large Hadron Collider, independently found tentative evidence for a Higgs boson with a mass of about 125 GeV (gigaelectronvolts, units used interchangeably for mass and energy in particle physics, since both are equivalent). On 4 July 2012 CERN, the European particle physics laboratory that operates the Large Hadron Collider, announced, to a packed audience of scientists and science journalists, that the continuum had crumbled in favour of the Higgs. Both groups had collected large amounts of additional data, and the chance that their data showed a random fluctuation, rather than a new particle with Higgs-like properties, had dropped below 1 in 2 million. This is the degree of confidence traditionally required in particle physics before breaking out the champagne.

Further experiments will be needed to make sure that the new particle has all of the features that a theoretical Higgs boson should possess. For example, theory predicts that the Higgs boson should have spin 0; at the time of the announcement, the observations showed it to be either 0 or 2. There is also a possibility that 'the' Higgs boson may turn out to be composed of other, smaller, particles, or that it is just the first in a new family of Higgs-like particles. So either the current model of fundamental particles will be cemented in place, or we will have new information that will eventually lead to a better theory.

The last of the seven millennium prize problems is closely related to the Standard Model and the Higgs boson. It is a central question in quantum field theory, the mathematical framework in which particle physics is studied. It is called the mass gap hypothesis, and it places a specific lower limit on the possible mass of a fundamental particle. It is one representative problem chosen from a series of big unsolved questions in this deep and very new area of mathematical physics. It has connections that range from the frontiers of pure mathematics to the long-sought unification of the two main physical theories, general relativity and quantum field theory.

In classical Newtonian mechanics, the basic physical quantities are space, time, and mass. Space is assumed to be three-dimensional Euclidean, time is a one-dimensional quantity independent of space, and mass signifies the presence of matter. Masses change their location in space under the influence of forces, and the rate at which their position changes is measured with respect to time. Newton's law of motion describes how a body's acceleration (the rate of change of velocity, which is itself the rate of change of position) relates to the mass of the body and the force applied.

The classical theories of space, time, and matter were brought to their peak in James Clerk Maxwell's equations for electromagnetism.[79] This elegant system of equations unified two of nature's forces, previously thought to be distinct. In place of electricity and magnetism, there was a single electromagnetic field. A field pervades the whole of space, as if the universe were filled with some kind of invisible fluid. At each point of space we can measure the strength and direction of the field, as if that fluid were flowing in mathematical patterns. For some purposes the electromagnetic field can be split into two components, the electric field and the magnetic field. But a moving magnetic field creates an electric one, and conversely, so when it comes to dynamics, both fields must be combined into a single more complex one.

This cosy picture of the physical world, in which the fundamental scientific concepts bear a close resemblance to things that our senses perceive, changed dramatically in the early years of the twentieth century. At that point, physicists began to realise that on very small scales, much too small to be observed in any microscope then available, matter is very different from what everyone had imagined. Physicists and chemists started to take seriously a rather wild theory that went back more than two millennia to the philosophical musings of Democritus in ancient Greece and other scholars in India. This was the idea that although the world seems to be made of countless different materials, all matter is built from tiny particles: atoms. The word comes from the Greek for 'indivisible'.

The nineteenth-century chemists found indirect evidence for atoms: the elements that combine together to form more complex molecules do so in very specific proportions, often close to whole numbers. John Dalton formulated these observations as his law of multiple proportions, and put forward atoms as an explanation. If each chemical compound consisted of fixed numbers of atoms of various kinds, this kind of ratio would automatically appear. For example, we now know that each molecule of carbon dioxide consists of two oxygen atoms and one carbon atom, so the numbers of atoms will be in the ratio two to one. However, there are complications: different atoms have different masses, and many elements occur as molecules formed from several atoms – for example, the oxygen molecule is composed of two oxygen atoms. If you don't realise what's going on, you will think that an atom of oxygen is twice as massive as it actually is. And some apparent elements are actually mixtures of different 'isotopes' – atomic structures. For instance, chlorine occurs in nature as a mixture of two stable forms, now called chlorine-35 and chlorine-

37, in proportions of about 76 per cent and 24 per cent respectively. So the observed 'atomic weight' is 35.45, which in the fledgling stages of atomic theory was interpreted as 'the chlorine atom is composed of thirty-five and a half hydrogen atoms'. And that means that an atom is not indivisible. As the twentieth century opened, most scientists still felt that the leap to atomic theory was too great, and the numerical evidence was too weak to justify taking it.

Some scientists, notably Maxwell and Ludwig Boltzmann, were ahead of the curve, convinced that gases are thinly distributed collections of molecules and that molecules are made by assembling atoms. What convinced most of their fellows seems to have been Albert Einstein's explanation of Brownian motion, erratic movements of tiny particles suspended in a fluid, that were visible under a microscope. Einstein decided that these movements must be caused by collisions with randomly moving molecules of the fluid, and he carried out some quantitative calculations to support that view. Jean Perrin confirmed these predictions experimentally in 1908. Being able to see the effect of the alleged indivisible particles of matter, and make quantitative predictions, carried more conviction than philosophical musings and curious numerology. In 1911 Amedeo Avogadro sorted out the problem with isotopes, and the existence of atoms became the scientific consensus.

While this was going on, a few scientists started to realise that atoms are not indivisible. They have some kind of structure, and it is possible to knock small pieces off them. In 1897 Joseph John Thomson was experimenting with so-called cathode rays, and he discovered that atoms could be made to emit even tinier particles, electrons. Not only that: the atoms of different elements emitted the same particles. By applying a magnetic field, Thomson showed that electrons carry a negative electric charge. Since an atom is electrically neutral, there must also be some part of atoms with a positive charge, leading Thomson to propose the plum pudding model: an atom is like a positively charged pudding riddled with negatively charged plums. But in 1909 one of Thomson's ex-students, Ernest Rutherford, performed experiments showing that most of the mass of an atom is concentrated near its centre. Puddings aren't like that.

How can experiments probe such tiny regions of space? Imagine a plot of land, which may or may not have buildings or other structures on it. You're not permitted to enter the area, and it's pitch dark so you can't see what's there. However, you do have a rifle and many boxes of ammunition. You can shoot bullets at random into the plot, and observe the direction in which they exit. If the plot is like a plum pudding, most bullets will go straight through. If you occasionally have to duck as a bullet ricochets straight back at you, there's something pretty solid somewhere. By observing how frequently the bullet exits at a given angle, you can estimate the size of the solid object.

Rutherford's bullets were alpha particles, nuclei of helium atoms, and his plot of land was a thin sheet of gold foil. Thomson's work had shown that the electron plums have very little mass, so almost all of the mass of an atom should be found in the pudding. If the pudding didn't have lumps, most alpha particles ought to go straight through, with very few being deflected, and then not by much. Instead, a small but significant proportion experienced large deflections. So the plum pudding picture didn't work. Rutherford suggested a different metaphor, one we still use informally today despite it being superseded by more modern images: the planetary model. An atom is like the solar system; it has a huge central nucleus, its sun, around which electrons orbit like planets. So, like the solar system, the interior of an atom is mostly empty space.

Rutherford went on to find evidence that the nucleus is composed of two distinct types of particle: protons, with positive charge; and neutrons, with zero charge. The two have very similar masses, and both are about 1800 times as massive as an electron. Atoms, far from being indivisible, are made from even smaller subatomic particles. This theory explains the integer numerology of chemical elements: what are being counted are the numbers of protons and neutrons. It also explains isotopes: adding or subtracting a few neutrons changes the mass, but keeps the charge zero and leaves the number of electrons – equal to the number of protons – unchanged. An atom's chemical properties are mostly controlled by its electrons. For instance, chlorine-35 has 17 protons, 17 electrons, and 18 neutrons; chlorine-37 has 17 protons, 17 electrons, and 20 neutrons. The 35.45 figure arises because natural chlorine is a mixture of these two isotopes.

By the early twentieth century there was a new theory in town, applicable to matter on the scales of subatomic particles. It was quantum mechanics, and once it became available, physics would never be the same. Quantum mechanics predicted a host of new phenomena, many of which were quickly observed in the laboratory. It explained a great many strange and previously baffling observations. It predicted the existence of new fundamental particles. And it told us that the classical image of the universe we live in, despite its previously excellent agreement with observations, is wrong. Our human-scale perceptions are poor models of reality at its most fundamental level.

In classical physics, matter is made from particles and light is a wave. In quantum mechanics, light is also a particle, the photon; conversely, matter, for instance electrons, can sometimes behave like a wave. The previously sharp dividing line between waves and particles did not so much become blurred as vanish altogether, replaced by wave/particle duality. The planetary model of the atom didn't work very well if you took it literally, so a new image arose. Instead of orbiting the nucleus like planets, electrons form a fuzzy cloud centred on the nucleus, a cloud not of stuff but of probabilities. The density of the cloud corresponds to the likelihood of finding an electron in that location.

As well as protons, neutrons, and electrons, physicists knew one further subatomic particle, the photon. Soon others appeared. An apparent failure of the law of conservation of energy led Wolfgang Pauli to propose patching it up by postulating the existence of the neutrino, an invisible and virtually undetectable new particle that would provide the missing energy. It was just detectable enough for its existence to be confirmed in 1956. And that opened the floodgates. Soon there were pions, muons, and kaons, the latter discovered by observing cosmic rays. Particle physics was born, and it continued to use Rutherford's method to probe the incredibly tiny spatial scales involved: to find out what's inside something, throw a lot of stuff at it and observe what bounces off. Ever larger particle accelerators – in effect, the guns that shot the bullets – were built and operated. The Stanford linear accelerator was 3 kilometres long. To avoid having to build accelerators whose lengths would span continents, they were bent into a circle, so particles could go round and round huge numbers of times at colossal speeds. That complicated the technology, because particles moving in circles radiate energy, but there were fixes.

The first fruit of these labours was an ever-growing catalogue of allegedly fundamental particles. Enrico Fermi expressed his frustration: 'If I could remember the names of all these particles, I'd be a botanist.' But every so often, new ideas from quantum theory collapsed the list again, as new kinds of ever-smaller particles were proposed to unify the structures already observed.

Early quantum mechanics applied to individual wave-like or particle-like things. But initially, no one could describe a good quantum-mechanical analogue of a field. It was impossible to ignore this gap because particles (describable by quantum mechanics) could and did interact with fields (not describable by quantum mechanics). It was like wanting to find out how the planets of the solar system move, when you knew Newton's laws of motion (how masses move when forces are applied), but didn't know his law of gravity (what those forces are).

There was another reason to want to model fields, rather than just particles. Thanks to wave/particle duality, they are intimately related. A particle is essentially a bunched-up chunk of field. A field is a sea of tightly packed particles. The two concepts are inseparable. Unfortunately, the methods developed to that date relied on particles being like tiny points, and they didn't extend in any sensible way to fields. You couldn't just stick lots of particles together and call the result a field, because particles *interact* with each other.

Imagine a crowd of people in ... well, a field. Perhaps they are at a rock concert. Viewed from a passing helicopter, the crowd resembles a fluid, sloshing around the field – often literally, for example at the Glastonbury Festival, renowned for becoming a sea of mud. Down on the ground, it becomes clear that the fluid is really a seething mass of individual particles: people. Or perhaps dense clusters of people, as a few friends walk close together, forming an indivisible unit, or as a group of strangers comes together with a common purpose, such as getting to the bar. But you can't model the crowd accurately by adding together what the people would do if they were on their own. As one group heads for the bar, it blocks the path of another group. The two groups collide and jostle. Setting up an effective quantum field theory is like doing this when the people are localised quantum wavefunctions.

By the end of the 1920s this kind of reasoning had convinced physicists that however hard the task might be, quantum mechanics had to be extended to take care of fields as well as particles. The natural place to start was the electromagnetic field. Somehow the electrical and magnetic components of this field had to be quantised: rewritten in the formalism of quantum mechanics. Mathematically, that formalism was unfamiliar and not terribly physical. Observables – things you could measure – were no longer represented using good old numbers. Instead, they corresponded to operators on a Hilbert space: mathematical rules for manipulating waves. These operators violated the usual assumptions of classical mechanics. If you multiply two numbers together, the result is the same whichever comes first. For instance, 2×3 and 3×2 are the same. This property, called commutativity, fails for many pairs of operators, much as putting on your socks and then your shoes does not have the same effect as putting on your shoes and then your socks. Numbers are passive creatures, operators are active. Which action you take first sets the scene for the other.

Commutativity is a very pleasant mathematical property. Its absence is a bit of a nuisance, and this is just one of the reasons why quantising a field turns out to be tricky. Nonetheless, it can sometimes be done. The electromagnetic field was quantised in a series of stages, starting with Dirac's theory of the electron in 1928, and completed by Sin-Itiro Tomonaga, Julian Schwinger, Richard Feynman, and Freeman Dyson in the late 1940s and early 1950s. The resulting theory is known as quantum electrodynamics.

The point of view that was used there hinted at a method that might work more generally. The underlying idea went right back to Newton. When mathematicians attempted to solve the equations supplied by Newton's law, they discovered some useful general tricks, known as conservation laws. When a system of masses moves, some quantities remain unchanged. The most familiar is energy, which comes in two flavours, kinetic and potential. Kinetic energy is related to how fast the body moves, and potential energy is the work done by forces. When a rock is pushed off the edge of a cliff it trades potential energy, due to gravity, for kinetic energy; in ordinary language, it falls and speeds up. Other conserved quantities are momentum, which is mass times velocity, and angular momentum, which is related to the body's rate of spin. These conserved quantities relate the different variables used to describe the system, and therefore reduce their number. That helps when solving the equations, as we saw for the two-body problem in chapter 8.

By the 1900s the source of these conservation laws had been understood. Emmy Noether proved that every conserved quantity corresponds to a continuous group of symmetries of the equations. A symmetry is a mathematical transformation that leaves the equations unchanged, and all symmetries form a group, with the operation being 'do one transformation, then the other'. A continuous group is a group of symmetries defined by a single real number. For example, rotation about a given axis is a symmetry, and the angle of rotation can be any real number, so the rotations – through any angle – about a given axis form a continuous family. Here the associated conserved quantity is angular momentum. Similarly, momentum is the conserved quantity associated with the family of translations in a given direction. What about energy? That is the conserved quantity corresponding to time symmetries – the equations are the same at all instants of time.

When physicists tried to unify the basic forces of nature, they became convinced that symmetries were the key. The first such unification was Maxwell's, combining electricity and magnetism into a single electromagnetic field. Maxwell achieved this unification without considering symmetry, but it soon became clear that his equations possess a remarkable kind of symmetry that had not previously been noticed: gauge symmetry. And that looked like a strategic lever that might open up more general quantum field theories.

Rotations and translations are global symmetries: they apply uniformly across the whole of space and time. A rotation about some axis rotates every point in space through the same angle. Gauge symmetries are different: they are local symmetries, which can vary from point to point in space. In the case of electromagnetism, these local symmetries are changes of phase. A local oscillation of the electromagnetic field has both an amplitude (how big it is) and a phase (the time at which it reaches its peak). If you take a solution of Maxwell's field equations and change the phase at each point, you get another solution, provided you make a compensating change to the description of the field, incorporating a local electromagnetic charge.

Gauge symmetries were introduced by Hermann Weyl in an abortive attempt to bring about a further unification, of electromagnetism and general relativity. That is, the electromagnetic and gravitational forces. The name came about because of a misapprehension: he thought that the correct local symmetries should be changes of spatial scale, or 'gauge'. This idea didn't work out, but the formalism of quantum mechanics led Vladimir Fock and Fritz London to introduce a different type of local symmetry. Quantum mechanics is formulated using complex numbers, not just real numbers, and every quantum wavefunction has a complex phase. The relevant local symmetries rotate the phase through any angle in the complex plane. Abstractly this group of symmetries consists of all rotations, but in complex coordinates these are 'unitary transformations' (U) in a space with one complex dimension (1), so the group formed by these symmetries is denoted by $U(1)$. The formalism here is not just an abstract mathematical game: it allowed physicists to write down, and then solve, the equations for charged quantum particles moving in an electromagnetic field. In the hands of Tomonaga, Schwinger, Feynman, and Dyson, this point of view led to the first relativistic quantum field theory of electromagnetism: quantum electrodynamics. Symmetry under the gauge group $U(1)$ was fundamental to their work.

The next step, unifying quantum electrodynamics with the weak nuclear force, was achieved by Abdus Salam, Sheldon Glashow, Steven Weinberg, and others in the 1960s. Alongside the electromagnetic field with its $U(1)$ gauge symmetry, they introduced fields associated with four fundamental particles, the so-called bosons W^+, W^0, W^-, and B^0. The gauge symmetries of this field, which in effect rotate combinations of these particles to produce other combinations, form another group, called $SU(2)$ – unitary (U) transformations in a two-dimensional complex space (2) that are also special (S), a simple technical condition. The combined gauge group is therefore $U(1) \times SU(2)$, where the × indicates that the two groups act independently on the two fields. The result, called the electroweak theory, required a difficult mathematical innovation. The group $U(1)$ for quantum electrodynamics is commutative: applying two symmetry transformations in turn gives the same result, whichever one is applied first. This pleasant property makes the mathematics much simpler, but it fails for SU(2). This was the first application of a non-commutative gauge theory.

The strong nuclear force comes into play when we consider the internal structure of particles like protons and neutrons. The big breakthrough in this area was motivated by a curious mathematical pattern in one particular class of particles, called hadrons. The pattern was known as the eightfold way. It inspired the theory of quantum chromodynamics, which postulated the existence of hidden particles called quarks, and used them as basic components for the large zoo of hadrons.

In the standard model, everything in the universe is made from sixteen genuinely fundamental particles, whose existence has been confirmed by accelerator experiments. Plus a seventeenth, for which the Large Hadron Collider is currently searching. Of the particles known to Rutherford, only two remain fundamental: the electron and the photon. The proton and the neutron, in contrast, are made from quarks. The name was coined by Murray Gell-Mann, who intended it to rhyme with 'cork'. He came across a passage in James Joyce's *Finnegans Wake*:

> Three quarks for Muster Mark!
> Sure he has not got much of a bark
> And sure any he has it's all beside the mark.

This would seem to hint at a pronunciation rhyming with 'mark', but Gell-Mann found a way to justify his intention. Both pronunciations are now common.

The Standard Model envisages six quarks, arranged in pairs. They have curious names: up/down, charmed/strange, and top/bottom. There are six leptons, also in pairs: the electron, muon, and tauon (today usually just called tau) and their associated neutrinos. These twelve particles collectively are called fermions, after Fermi. Particles are held together by forces, which are of four kinds: gravity, electromagnetism, the strong nuclear force, and the weak nuclear force. Leaving out gravity, which has not yet been fully reconciled with the quantum picture, this gives three forces. In particle physics, forces are produced by an exchange of particles, which 'carry' or 'mediate' the force. The usual analogy is two tennis players being held together by their mutual attention to the ball. The photon mediates the electromagnetic force, the Z- and W-bosons mediate the weak nuclear force, and the gluon mediates the strong nuclear force. Well, technically it mediates the colour force, which holds quarks together, and the strong force is what we observe as a result. The proton consists of two up quarks plus one down quark; the neutron consists of one up quark plus two down quarks. In each of these particles, the quarks are held together by gluons. These four force carriers are known collectively as bosons, after Chandra Bose. The distinction between fermions and bosons is important: they have different statistical properties. Figure 44 (left) shows the resulting catalogue of conjecturally fundamental particles. Figure 44 (right) shows how to make a proton and a neutron from quarks.

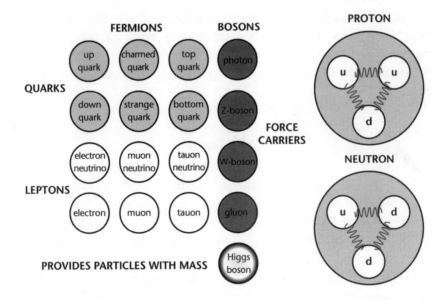

Fig 44 *Left:* The 17 particles of the Standard Model. *Right:* How to make a proton and a neutron from quarks. *Right top:* Proton = two up quarks + one down quark. *Right bottom:* Neutron = one up quark + two down quarks.

The Higgs boson completes this picture by explaining why the other 16 particles of the Standard Model have nonzero masses. It is named for Peter Higgs, one of the physicists who suggested the idea.

Others involved include Philip Anderson, François Englert, Robert Brout, Gerald Guralnik, Carl Hagen, and Thomas Kibble. The Higgs boson is the particle incarnation of a hypothetical quantum field, the Higgs field, with an unusual but vital feature: in a vacuum, the field is nonzero. The other 16 particles are influenced by the Higgs field, which makes them behave as if they have mass.

In 1993 David Miller, responding to a challenge from the British science minister William Waldegrave, presented a striking analogy: a cocktail party. People are spread uniformly around the room when the guest of honour (an ex–prime minister) walks in. Immediately everyone bunches up around her. As she moves across the room, different people join and leave the bunch, and the moving bunch gives her extra mass, making her harder to stop. This is the Higgs mechanism. Now imagine a rumour passing through the room, with people clustering together to hear the news. This cluster is the Higgs boson. Miller added: 'There could be a Higgs mechanism, and a Higgs field throughout our Universe, without there being a Higgs boson. The next generation of colliders will sort this out.' It now seems to have sorted out the Higgs boson, but the Higgs field still needs further work.

Quantum chromodynamics is another gauge theory, this time with gauge group SU(3). As the notation suggests, the transformations now act on a three-dimensional complex space. The unification of electromagnetism, the weak force, and the strong force then followed. It assumes the existence of three quantum fields, one for each force, with gauge groups U(1), SU(2), and SU(3) respectively. Combining all three yields the Standard Model, with gauge group U(1) × SU(2) × SU(3). Strictly speaking, the SU(2) and SU(3) symmetries are approximate; they are believed to become exact at very high energies. So their effect on the particles that make up our world correspond to 'broken' symmetries – the traces of structure that remain when the ideal perfectly symmetric system is subjected to small perturbations.

All three groups contain continuous families of symmetries: one such family of U(1), three for SU(2), and eight for SU(3). Associated with these are various conserved quantities. The symmetries of Newtonian mechanics again provide energy, momentum, and angular momentum. The conserved quantities for the U(1) × SU(2) × SU(3) gauge symmetries are various 'quantum numbers', which characterise particles. These are analogous to such quantities as spin and charge, but apply to quarks; they have names like colour charge, isospin, and hypercharge. Finally, there are some additional conserved quantities for U(1): quantum numbers for the six leptons, such as electron number, muon number, and tau number. The upshot is that the symmetries of the equations of the Standard Model, via Noether's theorem, explain all of the core physical variables of fundamental particles.

The important message for our story is the overall strategy and outcome. To unify physical theories, find their symmetries and unify those. Then devise a suitable theory with that combined group of symmetries. I'm not suggesting that the process is straightforward; it is actually technically very complex. But so far, this is how quantum field theory has developed, and only one of the four forces of nature currently falls outside its scope: gravity.

Not only does Noether's theorem explain the main physical variables associated with fundamental particles: that was how many of the underlying symmetries were found. Physicists worked backwards from observed and inferred quantum numbers to deduce what symmetries the model ought to have. Then they wrote down suitable equations with those symmetries, and confirmed that these equations fitted reality very closely. At the moment, this final step requires choosing the values of 19 parameters – numbers that must be plugged into the equations to provide quantitative results. Nine of these are masses of specific particles: all six quarks and the electron, muon, and tau. The rest are more technical, things like mixing angles and phase couplings. Seventeen of these parameters are known from experiments, but two are not; they describe the still-hypothetical Higgs *field*. But now there is a good prospect of measuring them, because physicists know where to look.

The equations employed in these theories belong to a general class of gauge field theories, known as Yang-Mills theories. In 1954 Chen-Ning Yang and Robert Mills attempted to develop gauge theories to explain the strong force and the particles associated with it. Their first attempts ran into trouble when the field was quantised, because this required the particles to have zero mass. In 1960 Jeffrey Goldstone,

Yoichiro Nambu, and Giovanni Jona-Lasinio found a way round this problem: start with a theory that predicted massless particles, but then modify it by breaking some of the symmetries. That is, change the equations a little by introducing new asymmetric terms. When this idea was used to modify Yang-Mills theory, the resulting equations performed very well both in the electroweak theory and in quantum chromodynamics.

Yang and Mills assumed the gauge group was a special unitary group. For the particle applications this was either SU(2) or SU(3), the special unitary group for two or three complex dimensions, but the formalism worked for any number of dimensions. Their theory tackles head-on a difficult but unavoidable mathematical difficulty. The electromagnetic field is in one respect misleadingly simple: its gauge symmetries commute. Unlike most quantum operators, the order in which you change phases doesn't affect the equations. What physicists had their eyes on was a quantum field theory for subatomic particles. There, the gauge group was non-commutative, which made quantising the equations very difficult.

Yang and Mills succeeded by using a diagrammatic representation of particle interactions introduced by Richard Feynman. Any quantum state can be thought of as a superposition of innumerable particle interactions. For example, even a vacuum involves pairs of particles and antiparticles momentarily winking into existence and then winking out again. A simple collision between two particles splits up into a bewildering dance of temporary appearances and disappearances of intermediary particles, shuttling back and forth, splitting and combining. What saves the day is a combination of two things. The field equations can be quantised for each specific Feynman diagram, and all of these contributions can be added together to represent the effect of the full interaction. Moreover, the most complicated diagrams are rare, so they don't contribute a lot to the sum. Even so, there is a serious problem. The sum, interpreted straightforwardly, is infinite. Yang and Mills found a way to 'renormalise' the calculation so that an infinity of terms that shouldn't really matter were removed. What was left was a finite sum, and its value matched reality very closely. This technique was totally mysterious when first invented, but it now makes sense.

In the 1970s, mathematicians got in on the act and Michael Atiyah generalised Yang-Mills theory to a broad class of gauge groups. The mathematics and physics began to feed off each other, and Edward Witten and Nathan Seiberg's work on topological quantum field theories led to the concept of supersymmetry, in which all known particles have new 'supersymmetric' counterparts: electrons and selectrons, quarks and squarks. This simplified the mathematics and led to physical predictions. However, these new particles have not yet been observed, and some probably should have shown up by now in the experiments carried out using the Large Hadron Collider. The mathematical value of these ideas is well established, but their direct relevance to physics is not. However, they shed useful light on Yang-Mills theory.

Quantum field theory is one of the fastest-moving frontiers of mathematical physics, so the Clay Institute wanted to include something on the topic as one of its millennium prizes. The mass gap hypothesis sits squarely in this rich area, and it addresses an important mathematical issue linked to particle physics. The application of Yang-Mills fields to describe fundamental particles in terms of the strong nuclear force depends on a specific quantum-theoretic feature known as a mass gap. In relativity, a particle that travels at the speed of light acquires infinite mass, unless its mass is zero. The mass gap allows quantum particles to have finite nonzero masses, even though the associated classical waves travel with the speed of light. When a mass gap exists, any state that is not the vacuum has an energy that exceeds that of the vacuum by at least some fixed amount. That is, there is a nonzero lower limit to the mass of a particle.

Experiments confirm the existence of a mass gap, and computer simulations of the equations support the mass gap hypothesis. However, we can't assume that a model matches reality and then use reality to verify mathematical features of the model, because the logic becomes circular. So theoretical understanding is needed. One key step would be a rigorous proof that quantum versions of Yang-Mills theory exist. The classical (non-quantum) version is fairly well understood nowadays, but the quantum analogue is bedevilled by the problem of renormalisation – those pesky infinities that have to be spirited away by mathematical trickery.

One attractive approach begins by turning continuous space into a discrete lattice and writing down a lattice analogue of the Yang-Mills equation. The main issue is then to show that as the lattice becomes increasingly fine, approximating a continuum, this analogue converges to a well-defined mathematical object. Some necessary features of the mathematics can be inferred from physical intuition, and it would be possible to prove that a suitable quantum Yang-Mills theory exists if these features could be established rigorously. The mass gap hypothesis involves a more detailed understanding of how the lattice theories approximate this hypothetical Yang-Mills theory. So existence of the theory, and the mass gap hypothesis, are closely intertwined.

That's where everyone is stuck. In 2004 Michael Douglas wrote a report on the status of the problem, and said: 'So far as I know, no breakthroughs have been made on this problem in the last few years. In particular, while progress has been made in lower dimensional field theories, I know of no significant progress towards a mathematically rigorous construction of the quantum Yang-Mills theory.' That assessment still seems to be correct.

Progress has been more impressive on some related problems, however, which may shed useful light. Special quantum field theories, known as two-dimensional sigma models, are more tractable, and the mass gap hypothesis has been established for one such model. Supersymmetric quantum field theories, involving hypothetical super-partners of the usual fundamental particles, have pleasant mathematical features which in effect remove the need for renormalisation. Physicists such as Edward Witten have been making progress towards solving related questions in the supersymmetric case. The hope is that some of the methods that emerge from this work might suggest new ways to tackle the original problem. But whatever the physical implications may be, and however the mass gap hypothesis eventually pans out, many of these developments have already enriched mathematics with important new concepts and tools.

14

Diophantine dreams
Birch–Swinnerton-Dyer Conjecture

I N CHAPTER 7 WE ENCOUNTERED the *Arithmetica* of Diophantus, and I remarked that six of its 13 books survive as Greek copies. Around AD 400, when ancient Greek civilisation went into decline, Arabia, China, and India took up the torch of mathematical innovation from Europe. Arabic scholars translated many of the classical Greek works, and often these translations are our main historical source for their contents. The Arab world knew about the *Arithmetica*, and built upon it. Four Arabic manuscripts discovered in 1968 may be translations of other 'missing' books from the *Arithmetica*.

Some time near the end of the tenth century AD, the Persian mathematician al-Karaji asked a question that could easily have occurred to Diophantus. Which integers can occur as the common difference between three rational squares that form an arithmetic sequence? For example the integer squares 1, 25, and 49 have common difference 24. That is, $1 + 24 = 25$ and $25 + 24 = 49$. Al-Karaji lived between about AD 953 and 1029, so he may have had access to a copy of the *Arithmetica*, but the earliest known translation was made by Abu'l-Wafā in 998. Leonard Dickson, who wrote a three-volume synopsis of the history of number theory, suggested that the problem might have originated some time before 972 in an anonymous Arab manuscript.

In algebraic language the problem becomes: for which integers d does there exist a rational number x such that $x - d$, x, and $x + d$ are all perfect squares? It can be restated in a form that is equivalent, though not obviously so: which whole numbers can be the area of a right-angled triangle with rational sides? That is: if a, b, and c are rational and $a^2 + b^2 = c^2$, what are the possible integer values for $ab/2$? Integers that satisfy these equivalent conditions are called congruent numbers. The term does not relate to other uses of the word 'congruent' in mathematics, and that makes it slightly confusing to a modern reader. Its origins are explained below.

Some numbers are not congruent: for example, it can be proved that 1, 2, 3, and 4 are not congruent. Others, such as 5, 6, and 7, are congruent. Indeed, the 3-4-5 triangle has area $3 \times 4/2 = 6$, proving that 6 is congruent. To prove 7 is congruent, observe that $(24/5)^2$, $(35/12)^2$, and $(337/60)^2$ have common difference 7. I'll come back to 5 in a moment. Proceeding case by case in this manner provides a lengthy list of congruent numbers, but sheds little light on their nature. No amount of case by case construction of examples can prove that a particular whole number is *not* congruent. For centuries no one knew whether 1 is congruent.

We now know that the problem goes far beyond anything that Diophantus could have solved. In fact, this deceptively simple question has still not been fully answered. The closest we've got is a characterisation of congruent numbers, discovered by Jerrold Tunnell in 1983. Tunnell's idea provides an algorithm for deciding whether a given integer can or cannot occur by counting its representations as two different combinations of squares. With a little ingenuity this calculation is feasible for quite large integers. The characterisation has only one serious disadvantage: it has never been proved correct. Its validity depends on solving one of the millennium problems, the Birch–Swinnerton-Dyer conjecture. This conjecture provides a criterion for an elliptic curve to have only finitely many rational points. We encountered these Diophantine equations in chapters 6 on the Mordell conjecture and 7 on Fermat's last theorem. Here we see further evidence for their prominent role at the frontiers of number theory.

The earliest European work referring to these questions was written by Leonardo of Pisa. Leonardo is best known for a sequence of strange numbers that he seems to have invented, which arose in an arithmetic problem about the progeny of some very unrealistic rabbits. These are the Fibonacci numbers

0 1 1 2 3 5 8 13 21 34 55 89 ...

in which each, after the first two, is the sum of the previous two. Leonardo's father was a customs official named Bonaccio, and the famous nickname means 'son of Bonaccio'. There is no evidence that it was used during Leonardo's lifetime, and it is thought to have been an invention of the French mathematician Guillaume Libri in the nineteenth century.[80] Be that as it may, Fibonacci numbers have many fascinating properties, and they are widely known. They even appear in Dan Brown's crypto-conspiracy thriller *The Da Vinci Code*.

Leonardo introduced the Fibonacci numbers in a textbook on arithmetic, the *Liber Abbaci* ('Book of calculation') of 1202, whose main aims were to draw European attention to the new arithmetical notation of the Arabs, based on the ten digits 0–9, and to demonstrate its utility. The idea had already reached Europe through al-Khwārizmī's text of 825 in its Latin translation *Algoritmi de Numero Indorum* ('On the calculation with Hindu numerals'), but Leonardo's book was the first to be written with the specific intention of promoting the uptake of decimal notation in Europe. Much of the book is devoted to practical arithmetic, especially currency exchange. But Leonardo wrote another book, not as well known, which in many ways was a European successor to Diophantus's *Arithmetica*: his *Liber Quadratorum* ('Book of squares').

Like Diophantus, he presented general techniques using special examples. One arose from al-Karaji's question. In 1225 Emperor Frederick II visited Pisa. He was aware of Leonardo's mathematical reputation, and seems to have decided that it would be fun to put it to the test in a mathematical tournament. Such public contests were common at the time. Contestants set each other questions. The emperor's team consisted of John of Palermo and Master Theodore. Leonardo's team consisted of Leonardo. The emperor's team challenged Leonardo to find a square which remains a square when 5 is added or subtracted. As usual, the numbers should be rational. In other words, they wanted a proof that 5 is a congruent number, by finding a specific rational x for which $x - 5$, x, and $x + 5$ are square.

This is by no means trivial – the smallest solution is

$$x = \frac{1681}{144} = \left(\frac{41}{12}\right)^2$$

in which case

$$x - 5 = \frac{961}{144} = \left(\frac{31}{12}\right)^2 \quad and \quad x + 5 = \frac{2401}{144} = \left(\frac{49}{12}\right)^2$$

Leonardo found a solution, and included it in the *Liber Quadratorum*. He got the answer using a general formula related to the Euclid/Diophantus formula for Pythagorean triples. From it, he obtained three integer squares with common difference 720, namely 31^2, 41^2, and 49^2. Then he divided by $12^2 = 144$ to get three squares with common difference 720/144, which is 5.[81] In terms of Pythagorean triples, take the 9, 40, 41 triangle with area 180 and divide by 36 to get a triangle with sides 20/3, 3/2, 41/6. Then its area is 5.

It is in Leonardo that we find the Latin word *congruum* for a set of three squares in arithmetic sequence. Later Euler used the word *congruere*, 'come together'. The first ten congruent numbers, and the corresponding simplest Pythagorean triples, are listed in Table 3. No simple patterns are evident.

d	Pythagorean triple
5	3/2, 20/3, 41/6
6	3, 4, 5
7	24/5, 35/12, 337/60
13	780/323, 323/30, 106921/9690
14	8/3, 63/6, 65/6
15	15/2, 4, 17/2
20	3, 40/3, 41/3
21	7/2, 12, 25/2
22	33/35, 140/3, 4901/105
23	80155/20748, 41496/3485, 905141617/72306780

Table 3 The first ten congruent numbers and corresponding Pythagorean triples.

Most of the early progress on this question was made by Islamic mathematicians, who showed that the numbers 5, 6, 14, 15, 21, 30, 34, 65, 70, 110, 154, and 190 are congruent, along with 18 larger numbers. To these, Leonardo, Angelo Genocchi (1855) and André Gérardin (1915) added 7, 22, 41, 69, 77, and 43 other numbers less than 1000. Leonardo stated in 1225 that 1 is not congruent, but gave no proof. In 1569 Fermat supplied one. By 1915 all congruent numbers less than 100 had been determined, but the problem yielded ground slowly, and by 1980 the status of many numbers less than 1000 remained unresolved. The difficulty can be judged by L. Bastien's discovery that 101 is congruent. The sides of the corresponding right-angled triangle are

$$\frac{711024064578955010000}{118171431852779451900}$$

$$\frac{3967272806033495003922}{118171431852779451900}$$

$$\frac{4030484925899520003922}{118171431852779451900}$$

He found these numbers in 1914, by hand. By 1986, with computers now on the scene, G. Kramarz had found all congruent numbers less than 2000.

At some point it was noticed that a different but related equation

$$y^2 = x^3 - d^2 x$$

has solutions x, y in whole numbers if and only if d is congruent.[82] This observation is obvious in one direction: the right-hand side is the product of x, $x - d$, and $x + d$, and if these are all squares, so is their product. The converse is fairly straightforward, too. This reformulation of the problem places it squarely within a rich and flourishing area of number theory. For any given d this equation sets y^2 equal to a cubic polynomial in x, and therefore defines an elliptic curve. So the problem of congruent numbers is a special case of a question that number theorists would dearly love to answer: when does an elliptic curve have at least one rational point? This question is far from straightforward, even for the special type of elliptic curve just mentioned. For instance, 157 is a congruent number, but the *simplest* right triangle with that area has hypotenuse

$$\frac{2244035177043369699245575130906674863160948472041}{8912332268928859588025535178967163570016480830}$$

Before proceeding further, we borrow Leonardo's trick, the one that led him from 720 to 5, and apply it in full generality. If we multiply any congruent number d by the square n^2 of an integer n, we also get a congruent number. Just take a rational Pythagorean triple corresponding to a triangle with area d, and multiply the numbers by n. The area of the triangle multiplies by n^2. The same is true if we divide the numbers by n; now the area divides by n^2. This process gives an integer only when the area has a square factor, so when seeking congruent numbers, it is enough to work with numbers that are squarefree – have no square factor. The first few squarefree numbers are

1 2 3 5 6 7 10 11 13 14 15 17 19

Now we can state Tunnell's criterion. An odd squarefree number d is congruent if and only if the number of (positive or negative) integer solutions x, y, z to the equation

$$2x^2 + y^2 + 8z^2 = d$$

is precisely twice the number of solutions to the equation

$$2x^2 + y^2 + 32z^2 = d$$

An even squarefree number d is congruent if and only if the number of integer solutions x, y, z to the equation

$$8x^2 + 2y^2 + 16z^2 = d$$

is precisely twice the number of solutions to the equation

$$8x^2 + 2y^2 + 64z^2 = d$$

These results are more useful than they might first seem. Because all the coefficients are positive, the sizes of x, y, z cannot exceed certain multiples of the square root of d. So the number of solutions is finite, and they can be found by a systematic search, with some useful short cuts. Here are the complete calculations for a few examples with small d:

- If $d = 1$ then the only solutions of the first equation are $x = 0$, $y = \pm 1$, $z = 0$. The same goes for the second equation. So both equations have two solutions, and the criterion fails to hold.

- If $d = 2$ then the only solutions of the first equation are $x = \pm 1$, $y = 0$, $z = 0$. The same goes for the second equation. So both equations have two solutions, and the criterion fails to hold.

- If $d = 3$ then the only solutions of the first equation are $x = \pm 1$, $y = \pm 1$, $z = 0$. The same goes for the second equation. So both equations have four solutions, and the criterion fails to hold.

- If $d = 5$ or 7 then the first equation has no solutions. The same goes for the second equation. Since twice zero is zero, the criterion is satisfied.

- If $d = 6$ we have to use the criterion for even numbers. Again both equations have no solutions, and the criterion is satisfied.

These simple calculations show that $1, 2, 3, 4 (= 2^2 \times 1)$ are not congruent, but 5, 6, and 7 are. The analysis can easily be extended, and in 2009 a team of mathematicians applied Tunnell's test to the first trillion numbers, finding exactly 3,148,379,694 congruent numbers. The researchers verified their results by performing the calculations twice, on different computers using different algorithms written by two independent groups. Bill Hart and Gonzalo Tornaria used the computer Selmer at the University of Warwick. Mark Watkins, David Harvey, and Robert Bradshaw used the computer Sage at the University of Washington.

However, there's a gap in all such calculations. Tunnell proved that if a number d is congruent, then it must satisfy his criterion. Therefore, if the criterion fails, the number is not congruent. This implies, for instance, that 1, 2, 3, and 4 are not congruent. However, he was unable to prove the converse: if a number satisfies his criterion, then it must be congruent. This is what we need to conclude that 5, 6, and 7 are congruent. In these particular cases we can find suitable Pythagorean triples, but that won't help with the general case. Tunnell did show that this converse follows from the Birch–Swinnerton-Dyer conjecture – but that remains unproved.

Like several of the millennium problems, the Birch–Swinnerton-Dyer conjecture is difficult even to state. (You think you can win a million dollars by doing something easy? I can sell you a lovely bridge, dead cheap ...) However, it rewards perseverance, because along the way we start to appreciate the depths, and long historical traditions, of number theory. If you look carefully at the name of the conjecture, one hyphen is longer than the other. It's not something conjectured by mathematicians named Birch, Swinnerton, and Dyer, but by Brian Birch and Peter Swinnerton-Dyer. Its full statement is technical, but it's about a basic issue in Diophantine equations – algebraic equations for which we seek solutions in whole or rational numbers. The question is simple: when do they have solutions?

In chapter 6 on Mordell's conjecture and chapter 7 on Fermat's last theorem we encountered some of the most wonderful gadgets in the whole of mathematics, elliptic curves. Mordell made what at the time was basically a wild guess, and conjectured that the number of rational solutions to an algebraic equation in two variables depends on the topology of the associated complex curve. If the genus is 0 – the curve is topologically a sphere – then the solutions are given by a formula. If the genus is 1 – the curve is topologically a torus, which is equivalent to it being an elliptic curve – then all rational solutions can be constructed from a suitable finite list by applying a natural group structure. If the genus is 2 or more – the curve is topologically a g-holed torus with $g \geqslant 2$ – then the number of solutions is finite. As we saw, Faltings proved this remarkable theorem in 1983.

The most striking feature of rational solutions to elliptic curve equations is that these solutions form a group, thanks to the geometric construction in Figure 28 of chapter 6. The resulting structure is called the Mordell-Weil group of the curve, and number theorists would dearly like to be able to calculate it. That involves finding a system of generators: rational solutions from which all others can be deduced by repeatedly using the group operation. Failing that, we would like at the very least to work out some of the basic features of the group, such as how big it is. Here, much detail is still not understood. Sometimes the group is infinite, so it leads to infinitely many rational solutions, sometimes it isn't, and the number of rational solutions is finite. It would be useful to be able to tell which is which. Indeed, what we would really like to know is the abstract structure of the group.

Mordell's proof that a finite list generates all solutions tells us that the group must be built from a finite group and a lattice group. A lattice group consists of all lists of integers of some fixed finite length. If the length is three, for instance, then the group consists of all lists (m_1, m_2, m_3) of integers, and lists are added in the obvious way:

$$(m_1, m_2, m_3) + (n_1, n_2, n_3) = (m_1 + n_1, m_2 + n_2, m_3 + n_3)$$

The length of the list is called the rank of the group (and geometrically it is the dimension of the lattice). If the rank is 0, the group is finite. If the rank is nonzero, the group is infinite. So to decide how many solutions there are, we don't need the full structure of the group. All we need is its rank. And that's what the Birch–Swinnerton-Dyer conjecture is about.

In the 1960s, when computers were just coming into being, the University of Cambridge had one of the earlier ones, called EDSAC. Which stands for electronic delay storage automatic calculator, and shows how proud its inventors were of its memory system, which sent sound waves along tubes of mercury and redirected them back to the beginning again. It was the size of a large truck, and I vividly remember being shown round it in 1963. Its circuits were based on thousands of valves – vacuum tubes. There were vast racks of the things along the walls, replacements to be inserted when a tube in the machine itself blew up. Which was fairly often.

Peter Swinnerton-Dyer was interested in the Diophantine side of elliptic curves, and in particular he wanted to understand how many solutions there would be if you replaced the curve by its analogue in a finite field with a prime number p of elements. That is, he wanted to study Gauss's trick of working 'modulo p'. He used the computer to calculate these numbers for lots of primes, and looked for interesting patterns.

Here's what he began to suspect. His supervisor John William Scott ('Ian') Cassels was highly sceptical at first, but as more and more data came in, he started to believe there might be something to the idea. What Swinnerton-Dyer's computer experiments suggested was this. Number theorists have a standard method which reinterprets any equation in ordinary integers in terms of integers to some modulus – recall 'clock arithmetic' to the modulus 12 in chapter 2. Because the rules of algebra apply in this version of arithmetic, any solution of the original equation becomes a solution of the 'reduced' equation to that modulus. Because the numbers involved form a finite list – only 12 numbers for clock arithmetic, for example – you can find all solutions by trial and error. In particular, you can count how many solutions there are, for any given modulus. Solutions to any modulus also impose conditions on the original integer solutions, and can sometimes even prove that such solutions exist. So it is a reflex among number theorists to reduce equations using various moduli, and primes are an especially useful choice.

So, to find out something about an elliptic curve, you can consider all primes up to some specific limit. For each prime, you can find how many points lie on the curve, modulo that prime. Birch noticed that Swinnerton-Dyer's computer experiments produce an interesting pattern if you divide the number of such points by the prime concerned. Then multiply all of these fractions together, for all primes less than or equal to a given one, and plot the results against successive primes on logarithmic graph paper. The data all seem to lie close to a straight line, whose slope is the rank of the elliptic curve. This led to a conjectured formula for the number of solutions associated with any prime modulus.[83]

The formula doesn't come from number theory, however: it involves complex analysis, the darling of the 1800s, which by some miracle is far more elegant than old-fashioned real analysis. In chapter 9 on the Riemann hypothesis we saw how analysis puts out tentacles in all directions, in particular having surprising and powerful connections with number theory. Swinnerton-Dyer's formula led to a more detailed conjecture about a type of complex function that I mentioned in chapter 9, called a Dirichlet *L*-function. This function is an analogue, for elliptic curves, of Riemann's notorious zeta function. The two mathematicians were definitely pushing the boat out, because at that time it wasn't known for sure that all elliptic curves *had* Dirichlet *L*-functions. It was a wild guess supported by the most tenuous of evidence. But as knowledge of the area grew, it came to appear ever more inspired. It wasn't a wild leap into the unknown: it was a wonderfully accurate, far-sighted stroke of refined mathematical intuition. Instead of standing on the shoulders of giants, Birch and Swinnerton-Dyer had stood on their own shoulders – giants who could hover in mid-air.

A basic tool in complex analysis is to express a function using a power series, like a polynomial but containing infinitely many terms, using bigger and bigger powers of the variable, which in this area is traditionally called s. To find out what a function does near some specific point, say 1, you use powers of $(s - 1)$. The Birch–Swinnerton-Dyer conjecture states that if the power series expansion near 1 of a Dirichlet L-function looks like

$$L(C, s) = c(s - 1)^r + \text{higher-order terms}$$

where c is a nonzero constant, then the rank of the curve is r, and conversely. In the language of complex analysis, this statement takes the form '$L(C, s)$ has a zero of order r at $s = 1$.'

The crucial point here is not the precise expression required: it is that given any elliptic curve, there exists an analytic calculation, using a related complex function, that tells us precisely how many independent rational solutions we have to find to specify them all.

Perhaps the simplest way to demonstrate that the Birch–Swinnerton-Dyer conjecture has genuine content is to observe that the largest known rank is 28. That is, there exists an elliptic curve having a set of 28 rational solutions, from which all rational solutions can be deduced. Moreover, no smaller set of rational solutions does that. Although curves of this rank are known to exist, no explicit examples have been found. The largest rank known for an explicit example is 18. The curve, found by Noam Elkies in 2006, is:

$$y^2 + xy = x^3 - 26175960092705884096311701787701203903556438$$
$$969515x + 51069381476131486489742177100373772089$$
$$79103253890567848326$$

As it stands this is not in the standard '$y^2 =$ cubic in x' form, but it can be transformed into that form at the expense of making the numbers even bigger. It is thought that the rank can be arbitrarily large, but this has not been proved. For all we know, the rank can never exceed some fixed size.

Most of what we can prove concerns curves of rank 0 or 1. When the rank is 0, there are finitely many rational solutions. When it is 1, then one specific solution leads to almost all of the rest, with perhaps a finite number of exceptions. These two cases include all elliptic curves of the form $y^2 = x^3 + px$ when p is a prime of the form $8k + 5$ (such as 13, 29, 37, and so on). It is conjectured that in these cases the rank is always 1, which implies that there are infinitely many rational solutions. Andrew Bremner and Cassels have proved this to be true for all such primes up to 1000. It can be tricky to find solutions that lead to almost all others, even when the rank is known, and small. They found that when $p = 877$ the *simplest* solution of this kind is the rational number

$$\frac{375494528127162193105504069942092792346201}{6215987776871505425463220780697238044100}$$

A great many theorems related to the Birch–Swinnerton-Dyer conjective have been proved, usually with very technical assumptions, but progress towards a solution has been relatively slight. In 1976 Coates and Wiles found the first hint that the conjecture might be true. They proved that a special kind of elliptic curve has rank 0 if the Dirichlet L-function does not vanish at 1. For such an elliptic curve, the number of rational solutions to the Diophantine equation is finite, perhaps zero – and you can deduce that from the corresponding L-function. Since then there have been a number of technical advances, still mostly limited to ranks 0 and 1. In 1990 Victor Kolyvagin proved that the Birch–Swinnerton-Dyer conjecture is true for ranks 0 and 1.

More detailed conjectures, with plenty of computer support, relate the constant c in the Birch–Swinnerton-Dyer conjecture to various number-theoretic concepts. There are analogues – equally enigmatic – for algebraic number fields. It is also known, in a precise sense, that most elliptic curves have rank 0 or 1. In 2010 Manjul Bhargava and Arul Shankar announced they had proved that the average rank of an elliptic curve is at most 7/6. If this and a few other recently announced theorems hold up under scrutiny, the Birch–Swinnerton-Dyer conjecture is true for a nonzero proportion of all elliptic curves. However, these are the simplest ones and they don't really represent the curves with a richer structure: rank 2 or more. These are an almost total mystery.

Complex cycles
Hodge Conjecture

SOME AREAS OF MATHEMATICS can be related, fairly directly, to everyday events and concerns. We don't encounter the Navier-Stokes equation in our kitchens, but we all understand what fluids are and have a feel for how they flow. Some areas can be related to esoteric questions in frontier science: you may need a PhD in mathematical physics to understand quantum field theory, but analogies with electricity and magnetism, or semi-meaningful images like 'probability wave', go a long way. Some can be explained using pictures: the Poincaré conjecture is a good example. But some defy all of these methods for making difficult abstract concepts accessible.

The Hodge conjecture, stated by the Scottish geometer William Hodge in 1950, is one of them. It's not the proof that causes problems, because there isn't one. It's the statement. Here it is from the Clay Institute's website, in a slightly edited form:

> On any non-singular projective complex algebraic variety, any Hodge class is a rational linear combination of classes of algebraic cycles.

Clearly we have work to do. The only words that make immediate sense are 'on, any, is, a,' and 'of'. Others are familiar as words: 'variety, class, rational, cycle'. But the images these conjure up – choice in the supermarket, a roomful of school children, unemotional thinking, a device with two wheels and handlebars – are obviously not the meanings that the Clay Institute has in mind. The rest are even more evidently jargon. But not jargon for jargon's sake – complicated names for simple things. These are simple names for complicated things. There are no ready-made names for such concepts in ordinary language, so we borrow some and invent others.

Looking on the positive side, we have a real opportunity here (as in 'boy, do we have opportunities'). The Hodge conjecture is arguably more representative of real mathematics, as done by mathematicians of the twentieth and twenty-first centuries, than any other topic in this book. By approaching it in the right way, we gain valuable insight into just how conceptually advanced frontier mathematics really is. Compared to school mathematics, it's like Mount Everest beside a molehill.

Is it all just airy-fairy pretentious nonsense carried out in ivory towers, then? If no ordinary person can understand what it's about, why should anyone hand over good tax money to employ the people who think about such matters? Let me turn that round. Suppose any ordinary person *could* understand everything that mathematicians think about. Would you be happy handing over the tax money then? Aren't they being paid for their expertise? If everything were so easy and comprehensible that it made immediate sense to anyone pulled in at random from the street, what would be the point of having mathematicians? If everybody knew how to drain a central heating system and solder a joint, what would be the point of having plumbers?

I can't show you any killer app that relies on the Hodge conjecture. But I can explain its importance within mathematics. Modern mathematics is a unified whole, so any significant advance, in any core area, will eventually prove its worth in dollars and cents terms. We may not find it in our kitchen today, but tomorrow, who knows? Closely related mathematical concepts are already proving their worth in several areas of science, ranging from quantum physics and string theory to robots.

Sometimes practical applications of new mathematics appear almost instantly. Sometimes it takes centuries. In the latter case, it might seem more cost-effective to wait until the need for such results arises and then run a crash programme to develop them. All mathematical problems that don't have immediate, obvious uses should be put on the back burner until they do. However, if we did that we'd always be behind the game, as mathematics spent a few hundred years playing catch-up with the needs of applied science. And it might not be at all clear which idea we need. Would you be happy if no one started thinking about how to make bricks until you'd hired a builder to begin work on a house? The more original a mathematical concept is, the less likely it would be to emerge from a crash programme.

A better strategy is to let some parts of mathematics develop along their own lines, and stop expecting immediate payoff. Don't try to cherry-pick; allow the mathematical edifice to grow organically. Mathematicians are cheap: they don't need expensive equipment like particle physicists do (Large Hadron Collider: €7.5 billion and counting). They pay their way by teaching students. Allowing a few of them to work part time on the Hodge conjecture, if that's what grabs them, is hardly unreasonable.

I'm going to unpick the statement of the Hodge conjecture, word by word. The easiest concept is 'algebraic variety'. It is a natural consequence of Descartes's use of coordinates to link geometry to algebra, chapter 3. With its aid, the tiny toolkit of curves introduced by Euclid and his successors – line, circle, ellipse, parabola, hyperbola – became a bottomless cornucopia. A straight line, the basis of Euclidean geometry, is the set of points that satisfy a suitable algebraic equation, for example $y = 3x + 1$. Change 3 and 1 to other numbers, and you get other lines. Circles need quadratic equations, as do ellipses, parabolas, and hyperbolas. In principle, anything you can state geometrically can be reinterpreted algebraically, and conversely. Do coordinates make geometry obsolete, then? Do they make algebra obsolete? Why use two tools when each does the same job as the other?

In my toolbox in the garage I have a hammer and a large pair of pincers. The hammer's job is to knock nails into wood, the pincers' job is to pull them out again. In principle, though, I could bash the nails in using the pincers, and the hammer has a claw specifically intended for extracting nails. So why do I need both? Because the hammer is better at some jobs, and the pincers are better at others. It's the same with geometry and algebra: some ways of thinking are more natural using geometry, some are more natural using algebra. It's the link between them that matters. If geometric thinking gets stuck, switch to algebra. If algebraic thinking gets stuck, switch to geometry.

Coordinate geometry provides a new freedom to invent curves. Just write down an equation and look at its solutions. Unless you've chosen a silly equation like $x = x$, you should get a curve. (The equation $x = x$ has the entire plane as its solutions.) For instance, I might write down $x^3 + y^3 = 3xy$, whose solutions are drawn in Figure 45. This curve is the folium of Descartes, and you won't find it in Euclid. The range of new curves that anyone can invent is literally infinite.

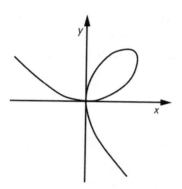

Fig 45 The folium of Descartes.

An automatic reflex among mathematicians is generalisation. Once someone has found an interesting idea, we can ask whether anything similar happens in a more general context. Descartes's idea has at least three major generalisations or modifications, all of which are needed to make sense of the Hodge conjecture.

First, what happens if we work with spaces other than the plane? Three-dimensional Euclidean space has three coordinates (x, y, z) instead of two. In space, one equation typically defines a surface. Two equations define a curve, where the corresponding surfaces meet. Three equations typically determine a point. (By 'typically' I mean that sometimes there might be exceptions, but these are very unusual and satisfy special conditions. We saw something similar in the plane with the silly equation $x = x$.)

Again, we can define new surfaces or curves, not found in Euclid, by writing down new equations. In the nineteenth century there was a vogue for doing this. You could publish a new surface if you said something genuinely interesting about it. A typical example is a surface introduced by Kummer in 1864, with the equation

$$x^4 + y^4 + z^4 - y^2z^2 - z^2x^2 - x^2y^2 - x^2 - y^2 - z^2 + 1 = 0$$

Figure 46 shows a picture. The main features of interest are the 16 'double points', where the shape is like two cones joined tip to tip. This is the maximum number possible for a quartic surface, one whose equation has degree 4, and that was interesting enough to merit publication.

Fig 46 Kummer's quartic surface with its 16 double points.

By the nineteenth century, mathematicians had experienced the heady delights of higher-dimensional spaces. There is no need to stop with three coordinates; why not try four, five, six, ... a million? This is not idle speculation. It is the algebra of lots of equations in lots of variables, and those turn up all over the mathematical landscape – for example in chapter 5 on the Kepler conjecture and chapter 8 on celestial mechanics. Nor was it idle generalisation: being able to think about such things geometrically, as well as algebraically, is a powerful tool that should not be restricted to spaces of two or three dimensions, just because that's where you can draw pictures and make models.

The word 'dimension' may sound impressive and mystical, but in this context it has a straightforward meaning: how many coordinates you need. For instance, 4-dimensional space has four coordinates (x, y, z, w), and as far as mathematics is concerned, that defines it. In four dimensions, a single equation typically defines a three-dimensional 'hypersurface', two equations define a surface (two dimensions), three equations define a curve (one dimension), and four equations define a point (zero dimensions). Each new equation gets rid of one dimension – one variable. So we can predict that in space of 17 dimensions, 11 equations define a 6-dimensional object, except for rare (and detectable) cases where some of the equations are superfluous.

An object defined in this way is called an algebraic variety. The word 'variety' arose in languages like French and Spanish, and has a similar meaning to 'manifold' in English: basically, the word 'many'. For reasons lost in the mists of history, 'manifold' became associated with topology and differential geometry – topology combined with calculus – while 'variety' became associated with algebraic geometry.[84] Using different names avoids confusion, so they both stuck. An algebraic variety could have been called a 'multidimensional space defined by a system of algebraic equations', but you can see why no one did that.

A second attractive way to generalise the notions of coordinate geometry is to allow the coordinates to be complex numbers. Recall that the complex number system involves a new kind of number, i, whose square is -1. Why complicate everything in that way? Because algebraic equations are much better behaved over the complex numbers. Over the real numbers, a quadratic equation may have two solutions or none. (It can also have just one, but there is a meaningful sense in which that is the same solution occurring twice.) Over the complex numbers, a quadratic equation *always* has two solutions (again counting multiplicities correctly). For some purposes, this is a far more pleasant property. You can say 'solve the equation for the seventh variable' and be confident that such a solution does actually exist.

Pleasant though it may be in this respect, complex algebraic geometry has features that take a little getting used to. With real variables, a line may cut a circle, or be tangent to it, or miss it altogether. With complex variables, the third option disappears. Once you get used to these changes, however, complex algebraic varieties are much better behaved than real ones. Sometimes real variables are essential, but for most purposes the complex context is a better choice. At any rate, we now know what a complex algebraic variety is.

How about 'projective'? This is the third generalisation, and it requires a slightly different notion of space. Projective geometry arose from Renaissance painters' interest in perspective, and it eliminates the exceptional behaviour of parallel lines. In Euclid's geometry, two straight lines either meet or they are parallel: they don't meet, no matter how far they extend. Now, imagine yourself standing on an infinite plane, paintbrush in hand, easel set up, paintbox at the ready, with a pair of parallel lines heading off towards the distant sunset like infinitely long railway lines. What do you see, and what would you draw? Not two lines that fail to meet. Instead, the lines appear to converge, meeting on the horizon.

What part of the plane does the horizon correspond to? The part where parallel lines meet. But there's no such thing. The horizon is the boundary, on your picture, of the image of the plane. If all's right with the world, that surely ought to be the image of the boundary of the plane. But a plane has no boundary. It goes on for ever. This is all a bit puzzling. It's as if part of the Euclidean plane is missing. If you 'project' a plane (the one with the railway lines) on to another plane (the canvas on the easel) you get a line in the image, the horizon, that isn't the projection of any line in the plane.

There is a way to get rid of this puzzling anomaly: add a so-called line at infinity to the Euclidean plane, representing the missing horizon. Now everything becomes much simpler. Two lines always meet at a point; the old notion of parallel lines corresponds to the case where two lines meet each other at infinity. This idea, suitably interpreted, can be turned into perfectly sensible mathematics. The result is called projective geometry. It's a very elegant subject, and the mathematicians of the eighteenth and nineteenth centuries loved it. Eventually they ran out of things to say, until the mathematicians of the twentieth century decided to generalise algebraic geometry to multidimensional spaces and to use complex numbers. At that point it became clear that we might as well go the whole hog and study complex solutions of systems of algebraic equations in projective space rather than real solutions in Euclidean space.

Let me sum up. A projective complex algebraic variety is like a curve, defined by an algebraic equation, but:

■ The number of equations and variables can be whatever we wish (algebraic variety).

■ The variables can be complex rather than real (complex).

■ The variables can take on infinite values in a sensible way (projective).

While we're at it, there's another term that can easily be dealt with: non-singular. It means that the variety is smooth, with no sharp ridges or places where the shape is more complicated than just a smooth piece of space. Kummer's surface is singular at those 16 double points. Of course we have to explain what 'smooth' means when the variables are complex and some can be infinite, but that's routine technique.

We're almost halfway along the statement of the Hodge conjecture. We know what we're talking about, but not how Hodge thought it ought to behave. Now we have to tackle the deepest and most technical aspects: algebraic cycles, classes, and (especially) Hodge classes. However, I can reveal the general gist straight away. They are technical gadgets that provide a partial answer to a very basic question about our generalised surface: *what shape is it?* The only remaining terms, 'rational linear combination', provide what everyone hopes is the right answer to that question.

See how far we've come. Already we understand what sort of statement the Hodge conjecture is. It tells us that given any generalised surface defined by some equations, you can work out what shape it is by doing some algebra with things called cycles. I could have told you that on the first page of this chapter, but at that stage it wouldn't have made any more sense than the formal statement did. Now that we know what a variety is, everything starts to hang together.

It also starts to sound like topology. 'Finding the shape by doing algebraic calculations' is strikingly reminiscent of Poincaré's ideas about algebraic invariants for topological spaces. So the next step requires a discussion of algebraic topology. Among Poincaré's discoveries were three important types of invariant, defined in terms of three concepts: homotopy, homology, and cohomology. The one we want is cohomology – and of course, wouldn't you just know it, that's the most difficult to explain.

I think we just have to jump in.

In three-dimensional space with real coordinates, a sphere and a plane meet (if they meet at all) in a circle. The sphere is a variety (I'll omit the adjective 'algebraic' when we speak of varieties), the circle is a variety, and the circle is contained in the sphere. We call it a *subvariety*. More generally, if we take the equations (many variables, complex, projective) that define some variety, and add some more equations, then we typically lose some of the solutions: those that fail to satisfy the new equations. The more equations we have, the smaller the variety becomes. The extended system of equations defines some part of the original variety, and this part is a variety in its own right – a subvariety.

When we count the number of solutions of a polynomial equation, it can be convenient to count the same point more than once. From this point of view, the set of solutions consists of a number of points, to each of which we 'attach' a number, its multiplicity. We might, for instance, have solutions 0, 1, and 2, with multiplicities 3, 7, and 4 respectively. The polynomial would then be $x^3(x-1)^7(x-2)^4$, if you want to know. Each of the three points $x = 0$, 1, or 2 is a (rather trivial) subvariety of the complex numbers. So the solutions of this polynomial can be described as a list of three subvarieties, with a whole number attached to each like a label.

An algebraic cycle is similar. Instead of single points, we use any finite list of subvarieties. To each of them we can attach a numerical label, which need not be a whole number. It could be a negative integer, it could be a rational number, it could be a real or even a complex number. For various reasons, the Hodge conjecture uses rational numbers as labels. This is what 'rational linear combination' refers to. So, for example, our original variety might be the unit sphere in 11-dimensional space, and this list might look like this:

A 7-dimensional hypersphere (with equations such and such) with label 22/7

A torus (with equations such and such) with label −4/5

A curve (with equations such and such) with label 413/6

Don't try to picture this, or if you do, think like a cartoonist: three squiggly blobs with little labels. Each such cartoon, each list, constitutes one algebraic cycle.

Why go to such fuss and bother to invent something so abstract? Because it captures essential aspects of the original algebraic variety. Algebraic geometers are borrowing a trick from topologists.

In chapter 10 on the Poincaré conjecture we thought about an ant whose universe is a surface. How can the ant work out what shape its universe is, when it can't pop outside and take a look? In particular, how can it distinguish a sphere from a torus? The solution presented there involved closed loops – topological bus trips. The ant pushes these loops around, finds out what happens when they are joined end to end, and computes an algebraic invariant of the space called its fundamental group. 'Invariant' means that topologically equivalent spaces have the same fundamental group. If the groups are different, so are the spaces. This is the invariant that led Poincaré to his conjecture. However, it's not easy for the poor ant to examine all possible loops in his universe, and this remark reflects genuine mathematical subtleties in the calculation of the fundamental group. A more practical invariant exists, and Poincaré investigated this as well. Pushing loops around is called homotopy. This alternative has a similar name: homology.

I'll show you the simplest, most concrete version of homology. Topologists quickly improved on this version, streamlined it, generalised it, and turned it into a huge mathematical machine called homological algebra. This simple version gives the barest flavour of how the topic goes, but that's all we need.

The ant starts by surveying its universe to make a map. Like a human surveyor, it covers its universe with a network of triangles. The crucial condition is that no triangle should surround a hole in the surface, and the way to ensure that is to create the triangles by slapping rubber patches on to the surface, like someone mending a bicycle tyre. Then each triangle has a well-defined interior that is topologically the same as the interior of an ordinary triangle in the plane. Topologists call such a patch a topological disc, because it is also equivalent to a circle and its interior. To see why, look at Figure 36 in chapter 10, where a triangle is deformed continuously into a circle. It's not possible to fit a patch of this kind to a triangle surrounding a hole, because the hole creates a tunnel that links the inside of the triangle to its outside. The patch would have to leave the surface, and the ant isn't allowed to do that.

The ant has now created a *triangulation* of its universe. The condition about patches ensures that the topology of the surface – its shape, in the sense of topological equivalence – can be reconstructed if all you know is the list of triangles, together with which triangles are adjacent to which. If you went to Ikea and bought an Ant Universe flatpack with suitably labelled triangles, and then glued edge A to edge AA, edge B to edge BB, and so on, you would be able to build the surface. The ant is confined to the surface, so it can't make a model, but it can be sure that in principle its map contains the information it needs. To extract that information, it has to perform a calculation. When doing so, the ant no longer has to contemplate the infinitude of all possible loops, but it does have to contemplate quite a lot of them: all closed loops that run along edges of its chosen network.

In homotopy, we ask whether a given loop can be shrunk continuously to a point. In homology, we ask a different question: does the loop form the boundary of a topological disc? That is, can you fit one or more triangular patches together so that the result is a region without any holes, and the boundary of this region is the loop concerned?

Figure 47 (left) shows part of a triangulation of a sphere, a closed loop, and the topological disc whose boundary it is. By setting up the right techniques, it can be proved that *any* loop in a triangulation of a sphere is a boundary: triangular patches, and more generally topological discs, are hole-detectors, and intuitively a sphere has no holes. However, a torus does have a hole, and indeed some loops on a torus are not boundaries. Figure 47 (right) shows such a loop, winding through the central hole. In other words: by running through a list of loops and finding out which of them are boundaries, the ant can distinguish a spherical universe from a toroidal one.

If the ant is as clever as Poincaré and the other topologists of his day, it can turn this idea into an elegant topological invariant, the homology group of its surface. The basic idea is to 'add' two loops by drawing both of them. However, that's not a loop, so we have to go back to the beginning and start again. Right back to the beginning, in fact; back to the days when we were first introduced to algebra. My mathematics

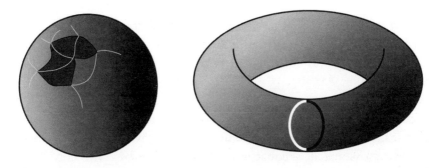

Fig 47 *Left*: Part of a triangulation of a sphere, a closed loop (black lines), and the disc whose boundary it is (dark shading). *Right*: Loop on a torus that is not the boundary of a disc (lighter part is at the back).

teacher started by pointing out that you can add a number of apples to a number of apples and get a total number of apples. But you can't add apples to oranges unless you count everything as fruit.

That's true in arithmetic, though even there you have to be careful not to use the same apple twice, but it's not true in algebra. There, you can add apples to oranges, while keeping them distinct. In fact, in advanced mathematics it is commonplace to add together things that you might imagine no one in their right mind would have invented at all, let alone want to add together. The freedom to do this kind of thing turns out to be amazingly useful and important, and the mathematicians who did it weren't mad after all – at least, not in that respect.

To understand some of the ideas that the Hodge conjecture brings together, we have to be able to add apples and oranges without lumping them all together as plain fruit. The way to add them is not actually very difficult. What's difficult is to accept that there's any point in doing so. Many of us have met a version of this potential conceptual block already. My teacher told the class that the letters stood for unknown numbers, with different letters for different unknowns. If you had a apples and another a apples, the total number of apples was $a + a = 2a$. And it worked whatever the number of apples might be. If you had $3a$ apples and added $2a$ apples, the result was $5a$, whatever the number of apples might be. The symbol, and what it represented, didn't matter either: if you had $3b$ oranges and added $2b$ oranges, the result was $5b$.[85] But what happened when you had $3a$ apples and $2b$ oranges? What was $3a$ plus $2b$?

$3a + 2b$.

That was it. You couldn't simplify the sum and make it 5 somethings – at least not without some manipulations involving a new category, fruit, and some new equations. That was the best you could do: live with it. However, once you took that step, you could do sums like

$$(3a + 2b) + (5a - b) = 8a + b$$

without introducing any new ideas. Or new kinds of fruit.

There were some caveats. I've already noted that if you add one apple to one apple, you only get two apples if the second apple is different from the first. The same goes for more complicated combinations of apples and oranges. Algebra assumes that for the purpose of doing the sums, all apples involved are different. In fact, it is often sensible to make this assumption, even in cases where two apples – or whatever else we are adding – might actually be the same. One apple plus the same apple is an apple with multiplicity two.

Once you get used to this idea, you can use it for anything. A pig plus the same pig is that pig with multiplicity two: pig + pig = 2 pig whatever pig is. A pig plus a cow is pig + cow. A triangle plus three circles is triangle + 3 circle. A superdupersphere plus three hyperelliptic quasiheaps is

superdupersphere + 3 hyperellipticquasiheap

whatever the jargon means (which, here, is nothing).

You can even allow negative numbers, and talk of three pigs minus eleven cows: 3 pig −11 cow. I have no idea what minus eleven cows look like, but I can be confident that if I add six cows to that, I've got minus five cows.[86] It's a formal game played with symbols, and no more realistic interpretation is needed, useful, or – often – possible. You could allow real numbers: π pigs minus $\sqrt{2}$ cows. Complex numbers. Any kind of fancy number that any mathematician has ever invented or will invent in future. The idea can be made a little more respectable if you think of the numbers as *labels*, attached to the pigs and cows. Now π pigs minus $\sqrt{2}$ cows can be thought of as a pig labelled π together with a cow labelled $-\sqrt{2}$. The arithmetic applies to the labels, not to the animals.

The Hodge conjecture involves this kind of construction, with extra bells and whistles. In place of animals, it uses curves, surfaces, and their higher-dimensional analogues. Strange as it may seem, the result is not just abstract nonsense, but a profound connection between topology, algebra, geometry, and analysis.

To set up the formalism of homology we want to add loops together, but not the way we did for the fundamental group. Instead, we do it the way my teacher told me. Just write the loops down and put a + sign in between. To make sense of that, we work not with single loops, but with finite sets of them. We label each loop with an integer that counts how often it occurs. Call such a labelled set a *cycle*. Now the ant can add any two cycles together by lumping them together and adding the corresponding labels, and the result is another cycle. Perhaps I should have used bicycles, not buses, as my image for the ant's travels in chapter 10.

When we were constructing the fundamental group, where 'addition' joins loops end to end, there was a technical snag. Adding the trivial loop to a loop didn't *quite* give the same loop, so the zero loop misbehaved. Adding a loop to its reversal didn't quite give the trivial loop, so inverses didn't behave correctly. The way out was to consider loops to be the same if one could be deformed into the other.

For homology, that's not the problem. There is a zero cycle (all labels zero), and every cycle does have an inverse (turn every label into its negative), so we do get a group. The trouble is, it's the wrong group. It tells us nothing about the topology of the space. To sort that out we use a similar trick, and take a more relaxed view of which cycles should count as zero. The ant cuts the space into triangular patches, and the boundary of each patch is topologically rather trivial: you can shrink it down to a point by pushing it all to the middle of its patch. So we require these boundary cycles to be equivalent to the zero cycle. It's a bit like turning ordinary numbers into clock arithmetic by pretending that the number 12 is irrelevant, so it can be set to zero. Here we turn cycles into homology by pretending that any boundary cycle is irrelevant.

The consequences of this pretence are dramatic. Now the algebra of cycles is affected by the topology of the space. The group of cycles modulo boundaries is a useful topological invariant, the homology group of the surface. At first sight it depends on which triangulation the ant chose, but as for the Euler characteristic, different triangulations of the same surface lead to the same homology group. So the ant has invented an algebraic invariant that can distinguish different surfaces. It's a bit fiddly, but you never get good invariants without doing some hard work somewhere along the line. This one is so effective that it can distinguish not just sphere from torus, but a 2-holed torus from a 5-holed torus, and similarly for any other numbers of holes.

Homology may seem a bit of a mouthful, but it started a rich vein of topological invariants, and it is based on simple geometric ideas: loops, boundaries, lumping sets together, doing arithmetic with labels. Considering that the poor ant is confined to its surface, it's astonishing that the creature can find out anything significant about the shape of its universe just by slapping down triangular patches, making a map, and doing some algebra.

There is a natural way to extend homology to higher dimensions. The 3-dimensional analogue of a triangle is a tetrahedron; it has 4 vertexes, 6 edges, 4 triangular faces, and a single 3-dimensional 'face', its interior. More generally, in n dimensions we can define an n-simplex with $n + 1$ vertexes, joined in pairs by all possible edges, which in turn form triangles, which assemble to create tetrahedrons, and so on. It is now easy to define cycles, boundaries, and homology, and again we can concoct a group by adding (homology classes of) cycles. In fact, we now get a whole series of groups: one for 0-dimensional cycles (points), one for 1-dimensional cycles (lines), one for 2-dimensional cycles (triangles), and so on, all the way up to the dimension of the space itself. These are the 0th, 1st, 2nd, and so on, homology groups of the space. Roughly speaking, they make precise the notion of holes, of various dimensions, in the space: do they exist, how many are there, and how do they relate to each other?

That, then, is homology, and it's almost what we need to

understand what the Hodge conjecture *says*. However, what we actually need is a closely related concept called *cohomology*. In 1893 Poincaré noticed a curious coincidence in the homology of any manifold: the list of homology groups reads the same in reverse. For a manifold of dimension 5, say, the 0th homology group is the same as the 5th, the 1st homology group is the same as the 4th, and the 2nd homology group is the same as the 3rd. He realised that this couldn't just be coincidence, and he explained it in terms of the dual of a triangulation, which we met in chapter 4 in connection with maps. This is a second triangulation in which each triangle is replaced by a vertex, each edge between two triangles by an edge that links the corresponding new vertexes, and each point by a triangle, as in Figure 9 of chapter 4. Notice how the dimensions appear in reverse order: 2-dimensional triangles become 0-dimensional points, and conversely; 1-dimensional edges remain 1-dimensional because 1 is in the middle.

It turns out to be useful to distinguish the two lists, even though they yield the same invariants. When the whole setup is generalised and formulated in abstract terms, triangulations disappear, and the dual triangulation no longer makes sense. What survives are two series of topological invariants, called homology groups and cohomology groups. Every concept in homology has a dual, usually named by adding 'co' at the front. So in place of cycle we have cocycles, and in place of two cycles being homologous we have two cocycles being cohomologous. The classes referred to in the Hodge conjecture are cohomology classes, and these are collections of cocycles that are cohomologous to each other.

Homology and cohomology don't tell us everything we would like to know about the shape of a topological space – distinct spaces can have the same homology and cohomology – but they do provide a lot of useful information, and a systematic framework in which to calculate it and use it.

An algebraic variety, be it real, complex, projective, or not, is a topological space. Therefore it has a shape. To find out useful things about the shape, we think like topologists and calculate the homology and cohomology groups. But the natural ingredients in algebraic geometry aren't geometric objects like triangulations and cycles. They are the things we can most easily describe by algebraic equations. Go back and look at the equation for Kummer's surface. How would that relate to a triangulation? There's nothing in the formula that hints at triangles.

Maybe we need to start again. Instead of triangles, we should use the natural building blocks for varieties, which are subvarieties, defined by imposing extra equations. Now we have to redefine cycles: instead of sets of triangles with integer labels, we use sets of subvarieties with whatever labels do the best job. For various reasons – mostly that the Hodge conjecture is false if we use integer labels – rational numbers are the sensible choice. Hodge's question boils down to this: does this new definition of homology and cohomology capture everything that the topological definition does? If his conjecture is true, then the algebraic cycle tool is sharp enough to match the cohomological chisel of topology. If it's false, then the algebraic cycle is a blunt instrument.

Except ... sorry, I've over-egged the pudding. The conjecture says that it is enough to use a particular *kind* of algebraic cycle, one that lives in a Hodge class. To explain that, we need yet another ingredient in an already rich mixture: analysis. One of the most important concepts in analysis is that of a differential equation, which is a condition about the rates at which variables change, chapter 8. Nearly all of the mathematical physics of the eighteenth, nineteenth, and twentieth centuries models nature using differential equations, and even in the twenty-first, most does. In the 1930s this idea led Hodge to a new body of technique, now called Hodge theory. It ties in naturally with a lot of other powerful methods in the general area of analysis and topology.

Hodge's idea was to use a differential equation to organise the cohomology classes into distinctive types. Each piece has extra structure, which can be used to advantage in topological problems. The pieces are defined using a differential equation that appeared in the late 1700s, notably in the work of Pierre-Simon de Laplace. Accordingly, it is called the Laplace equation. Laplace's main research was in celestial mechanics, the motion and form of planets, moons, comets, and stars. In 1783 he was working on the detailed shape of the Earth. By then it was known that the Earth is not a sphere, but it is flattened at the poles to form an oblate spheroid – like a beachball that someone is sitting on. But even that description misses some of the fine detail. Laplace found a method to calculate the shape to any required accuracy, based on a physical quantity that represents the Earth's gravitational field: not the field itself, but its gravitational potential. This is a measure of the energy contained in gravitation, a numerical quantity defined at each point in space. The force of gravity acts in whichever direction makes the potential decrease at the fastest rate, and the magnitude of the force is the rate of decrease.

The potential satisfies Laplace's equation: roughly speaking, this says that in the absence of matter – that is, in a vacuum – the average value of the potential over a very small sphere is equal to its value at the centre of the sphere. It's a kind of democracy: your value is the average of the values of your neighbours. Any solution of Laplace's equation is called a harmonic function. Hodge's special types of cohomology class are those that bear a particular relationship to harmonic functions. Hodge theory, the study of these types, opened up a deep and wonderful area of mathematics: relations between the topology of a space and a special differential equation on that space.

So now you have it. The Hodge conjecture postulates a deep and powerful connection between three of the pillars of modern mathematics: algebra, topology, and analysis. Take any variety. To understand its shape (topology, leading to cohomology classes) pick out special instances of these (analysis, leading to Hodge classes by way of differential equations). These special types of cohomology class can be realised using subvarieties (algebra: throw in some extra equations and look at algebraic cycles). That is, to solve the topology problem 'what shape is this thing?' for a variety, turn the question into analysis and then solve that using algebra.

Why is that important? The Hodge conjecture is a proposal to add two new tools to the algebraic geometer's toolbox: topological invariants and Laplace's equation. It's not really a conjecture about a mathematical theorem; it's a conjecture about new kinds of tools. If the conjecture is true, those tools immediately acquire new significance, and can potentially be used to answer an endless stream of questions. Of course, it might turn out to be false. That would be disappointing, but it's better to understand a tool's limitations than to keep hitting your thumb with it.

Now that we appreciate the nature of the Hodge conjecture, we can look at the evidence for it. What do we know? Precious little.

In 1924, before Hodge made his conjecture, Solomon Lefschetz proved a theorem which boils down to the Hodge conjecture for the dimension-2 cohomology of any variety. With a bit of routine algebraic topology, this implies the Hodge conjecture for varieties of dimension 1, 2, and 3. For higher-dimensional varieties, only a few special cases of the Hodge conjecture are known.

Hodge originally stated his conjecture in terms of integer labels. In 1961 Michael Atiyah and Friedrich Hirzebruch proved that in higher dimensions, this version of his conjecture is false. So today we interpret Hodge's conjecture using rational labels. For this version, there is a certain amount of encouraging evidence. The strongest evidence in its favour is that one of its deeper consequences, an even more technical theorem known as 'algebraicity of Hodge loci', has been proved – *without* assuming the Hodge conjecture. Eduardo Cattani, Pierre Deligne, and Aroldo Kaplan found such a proof in 1995.

Finally, there is an attractive conjecture in number theory that is analogous to the Hodge conjecture. It is called the Tate conjecture, after John Tate, and it links algebraic geometry to Galois theory, the circle of ideas that proved that there is no algebraic formula to solve polynomial equations of degree 5. Its formulation is technical, and it involves yet another version of cohomology. There are independent reasons to hope that the Tate conjecture might be true, but its status is currently open. But at least there is a sensible relative of the Hodge conjecture, even if it currently seems equally intractable.

The Hodge conjecture is one of those annoying mathematical assertions for which the evidence either for or against it is not very extensive, and not particularly convincing. There is a definite danger that the conjecture could be wrong. Perhaps there is a variety with a million dimensions that disproves the Hodge conjecture, for reasons that boil down to series of structureless calculations, so complicated that no one could ever carry them out. If so, the Hodge conjecture could be false for essentially silly reasons – it just happens not to be true – but virtually impossible to disprove. I know some algebraic geometers who suspect just that. If so, those million dollars will be safe for the foreseeable future.

16

Where next?

PREDICTION IS VERY DIFFICULT, especially about the future,[87] as Nobel-winning physicist Niels Bohr and baseball player and team manager Yogi Berra are supposed to have said.[88] Mind you, Berra also said: 'I never said most of the things I said.' Allegedly. Arthur C. Clarke, famous for his science fiction and the movie *2001: A Space Odyssey* and its sequels, was also a futurologist: he wrote books predicting the future of technology and society. Among the many predictions in his 1962 *Profiles of the Future* are:

Understanding the languages of whales and dolphins by 1970
Fusion power by 1990
Detection of gravity waves by 1990
Colonising planets by 2000

None of these has yet happened. On the other hand, he had some successes:

Planetary landings by 1980 (though he may have meant human landings)
Translating machines by 1970 (a bit premature, but they now exist on Google)
Personal radio by 1990 (mobile phones work like that)

He also predicted that we would have a global library by 2000, and this may be nearer the mark than we thought a few years ago, because this is one of the many functions of the Internet. With the advent of cloud computing, we may all end up using the same giant computer.

He missed some of the most important trends, such as the rise of the computer and genetic engineering, though he did predict this for 2030. With Clarke's uneven record as a warning, it would be foolhardy to predict the future of great mathematical problems in any detail. However, I can make some educated guesses, safe in the knowledge that most of them will turn out to be wrong.

In the introduction I mentioned Hilbert's 1900 list of 23 big problems. Most are now solved, and his brave war-cry 'We must know, we shall know' may seem vindicated. However, he also said 'in mathematics there is no *ignorabimus* [we shall be ignorant]' and Kurt Gödel knocked that idea firmly on the head with his incompleteness theorem: some mathematical problems may not have a solution within the usual logical framework of mathematics. Not just be impossible, like squaring the circle: they can be undecidable, meaning that no proof exists and no disproof exists. Possibly this could be the fate of some of the currently unsolved great problems. I'd be surprised if the Riemann hypothesis were like that, and amazed if anyone could prove it to be undecidable even if it were. On the other hand, the P/NP problem could well turn out to be undecidable, or to satisfy some other technical variation on the theme of 'it can't be done.' It has that kind of – well, *smell*.

I suspect that by the end of the twenty-first century we will have proofs of the Riemann hypothesis, the Birch–Swinnerton-Dyer conjecture, and the mass gap hypothesis, along with disproofs of the Hodge conjecture and the regularity of solutions of the Navier-Stokes equation in three dimensions. I expect P/NP still to be unsolved in 2100, but to succumb some time in the twenty-second century. So of course someone will disprove the Riemann hypothesis tomorrow and prove P is different from NP next week.

I'm on safer ground with general observations, because we can learn from history. So I'm reasonably confident that by the time the seven millennium problems have been solved, many of them will be seen as minor historical curiosities. 'Oh, they used to think *that* was important, did they?' This is what happened to some of the problems on Hilbert's hit-list. I can also be confident that within 50 years several major areas of mathematics that don't exist today will have come into being. It will then transpire that a few basic examples and some rudimentary theorems in these areas existed long before, but no one realised that these isolated snippets were clues to deep and important new areas. This is what happened with group theory, matrix algebra, fractals, and chaos. I don't doubt that it will happen again, because it's one of the standard ways in which mathematics develops.

These new areas will arise through two main factors. They will emerge from the internal structure of mathematics itself, or they will be responses to new questions about the outside world – often both in tandem. Like Poincaré's three-step process to problem-solving – preparation, incubation, and illumination – the relation between mathematics and its applications is not a single transition: science poses a problem, mathematics solves it, done. Instead, we find an intricate network of trade in questions and ideas, as new mathematics sparks further experiments or observations or theories, which in turn motivate new mathematics. And each node of this network turns out, on closer examination, to be a smaller network of the same kind.

There is more outside world than there used to be. Until recently, the main external source of inspiration for mathematics was the physical sciences. A few other areas played their part: biology and sociology influenced the development of probability and statistics, and philosophy had a big effect on mathematical logic. In the future we will see growing contributions from biology, medicine, computing, finance, economics, sociology, and very possibly politics, the movie industry, and sport. I suspect that some of the first new great problems will arise from biology, because that link is now firmly in place. One trend is an explosion in our ability to gather biological and biochemical data; small genomes can now be sequenced using a device the size of a memory stick, based on nanopore technology, for instance. Big genomes will rapidly follow using this or different technology, most of it already in existence.

These developments are potential game-changers, but we need to have better methods for understanding what the data imply. Biology isn't really about data as such. It is about processes. Evolution is a process, and so are the division of a cell, the growth of an embryo, the onset of cancer, the movement of a crowd, the workings of the brain, and the dynamics of the global ecosystem. The best way we currently know to take the basic ingredients of a process, and deduce what it does, is mathematics. So there will be great problems of new kinds – how dynamics unfolds in the presence of complex but specific organising information (DNA sequences); how genetic changes conspire with environment to constrain evolution; how rules for cell growth, division, mobility, stickiness, and death give developing organisms their shape; how the flow of electrons and chemicals in a network of nerve cells determines what it can perceive or how it will act.

Computing is another source of new mathematics that already has a track record. It is usually thought of as a tool for doing mathematics, but mathematics is equally a tool for understanding and structuring computations. This two-way trade is becoming increasingly important to the health and development of both areas, and they may even merge at some point in the future. Some mathematicians feel they should never have been allowed to split. Among the many trends visible here, the question of very large data sets again springs to mind. It relates not just to the DNA example mentioned earlier, but to earthquake prediction, evolution, the global climate, the stock market, international finance, and new technologies. The problem is to use large quantities of data to test and refine mathematical models of the real world, so that they give us genuine control over very complex systems.

The prediction about which I am most confident is in some ways negative, but it is also an affirmation of the continuing creativity of the mathematical community. All research mathematicians feel, from time to time, that their subject has a mind of its own. Problems work out the way mathematics wants them to, not how mathematicians want them to. We can choose what questions to ask, but we can't choose what answers we get. This feeling relates to two major schools of thought about the nature of mathematics. Platonists think that the 'ideal forms' of mathematics have some kind of independent existence, 'out there' in some realm distinct from the physical world. (There are subtler ways to say that which probably sound more sensible, but that's the gist.) Others see mathematics as a shared human construct. But unlike most such things – the legal system, money, ethics, morality – mathematics is a construct with a strong logical skeleton. There are severe constraints on what assertions you can or cannot share with everyone else. It is these constraints that give the impression that mathematics has its own agenda, and create the feeling in the minds of mathematicians that mathematics itself *exists* outside the domain of human activity. Platonism, I think, is not a description of what mathematics is. It is a description of what mathematics *feels like* when you are doing it. It is like the vivid sensation of 'red' that we experience when we see a rose, blood, or a traffic-light. Philosophers call these sensations qualia (singular: quale), and some of them think that our sensation of free will is actually a quale of the brain's way of making decisions. When we decide between alternatives, we feel that we have a genuine choice – whether or not the dynamic of the brain is actually deterministic in some sense. Similarly, Platonism is a quale of taking part in a shared human construct within a rigid framework of logical deduction.

So mathematics can seem to have a mind of its own, even though it is created by a collective of human minds. History tells us that the mathematical mind – in this sense – is more innovative and surprising than any single human mind can predict. All of which is a complicated way of getting to my main point: one thing we can safely predict about the future of mathematics is that it will be unpredictable. The most important mathematical questions of the next century will emerge as natural, perhaps even inevitable, consequences of our growing understanding of what we currently believe to be the great problems of mathematics. However, they will almost certainly be questions that we cannot currently conceive of. That is only right and proper, and we should celebrate it.

17

Twelve for the future

I don't want to leave you with the impression that most mathematical problems have been solved, aside from the odd really difficult one. Mathematical research is like exploring a new continent. As the area that we know about expands, the frontier that borders the unknown gets longer. I'm not suggesting that the more mathematics we discover, the less we know; I'm saying that the more mathematics we discover, the more we realise that we don't know. But what we don't know changes as time passes, with some old problems disappearing while new ones are added. In contrast, what we know just gets bigger – barring the occasional lost document.

To give you a tiny indication of what we currently *don't* know, aside from the great problems already discussed, here are twelve unsolved problems that have been baffling the world's mathematicians for quite a while. I've chosen them so that it is easy to understand the questions. As has amply been demonstrated, that carries no implications about how easy it may be to find the answers. Some of these problems may turn out to be great: that will mainly depend on the methods invented to solve them and what they lead to, not on the answer as such.

Brocard's Problem

For any whole number n, its factorial $n!$ is the product

$$n \times (n - 1) \times (n - 2) \times \cdots \times 3 \times 2 \times 1$$

This is the number of different ways to arrange n objects in order. For instance, the English alphabet with 26 letters can be arranged in

$$26! = 403,291,461,126,605,635,584,000,000$$

different orders. In articles written in 1876 and 1885, Henri Brocard noted that

$$4! + 1 = 24 + 1 = 25 = 5^2$$
$$5! + 1 = 120 + 1 = 121 = 11^2$$
$$7! + 1 = 5040 + 1 = 5041 = 71^2$$

are all perfect squares. He found no other factorials which became square when increased by 1, and asked whether any existed. The self-taught Indian genius Srinivasa Ramanujan independently asked the same question in 1913. Bruce Berndt and William Galway used a computer in 2000 to show that no further solutions exist for factorials of numbers up to 1 billion.

Odd Perfect Numbers
A number is perfect if it is equal to the sum of all of its proper divisors (that is, numbers that divide it exactly, excluding the number itself). Examples include

$$6 = 1 + 2 + 3$$
$$28 = 1 + 2 + 4 + 7 + 14$$

Euclid proved that if $2^n - 1$ is prime, then $2^{n-1}(2^n - 1)$ is perfect. The above examples correspond to $n = 2, 3$. Primes of this form are called Mersenne primes, and 47 of them are known, the largest to date being $2^{43,112,609} - 1$, also the largest known prime.[89] Euler proved that all even perfect numbers must be of this form, but no one has ever found an odd perfect number, or proved that they cannot exist. Pomerance has devised a non-rigorous argument which indicates that they don't. Any odd perfect number must satisfy several stringent conditions. It must be at least 10^{300}, it must have a prime factor greater than 10^8, its second largest prime factor must be at least 10^4, and it must have at least 75 prime factors and at least 12 distinct prime factors.

Collatz Conjecture

Take a whole number. If it is even, divide by 2. If it is odd, multiply by 3 and add 1. Repeat indefinitely. What happens?

For example, start with 12. Successive numbers are

$$12 \to 6 \to 3 \to 10 \to 5 \to 16 \to 8 \to 4 \to 2 \to 1$$

after which the sequence $4 \to 2 \to 1 \to 4 \to 2 \to 1$ repeats for ever. The Collatz conjecture states that the same end result occurs no matter which number you start with. It is named after Lothar Collatz who came up with it in 1937, but it has many other names: $3n + 1$ conjecture, hailstone problem, Ulam conjecture, Kakutani's problem, Thwaites conjecture, Hasse's algorithm, and Syracuse problem.

What makes the problem difficult is that the numbers can often explode. For instance, if we start with 27 then the sequence rises to 9232; even so, it finally gets down to 1 after 111 steps. Computer simulations verify the conjecture for all initial numbers up to 5.764×10^{18}. It has been proved that no cycles other than $4 \to 2 \to 1$ exist that involve fewer than 35,400 numbers. The possibility that some initial number leads to a sequence that contains ever-larger numbers, separated by smaller ones, has not been ruled out. Ilia Krasikov and Jeffrey Lagarias have proved that for initial values up to n, at least a contant times $n^{0.84}$ of them eventually get to 1. So exceptions, if they exist, are rare.[90]

Existence of Perfect Cuboids

This takes as its starting point the existence of, and formula for, Pythagorean triples, and it moves the problem into the third dimension. An Euler brick is a cuboid – a brick-shaped block – with integer sides, all of whose faces have integer diagonals. The smallest Euler brick was discovered in 1719 by Paul Halcke. Its edges are 240, 117, and 4; the face diagonals are 267, 244, and 125. Euler found formulas for such bricks, analogous to the formula for Pythagorean triples, but these do not give all solutions.

It is not known whether a perfect cuboid exists: that is, whether there is an Euler brick whose main diagonal, cutting through the interior of the brick from one corner to an opposite one, also has integer length. (There are four such diagonals but they all have the same length.) It is known that Euler's formulas cannot provide an example. Such a brick, if it exists, must satisfy several conditions – for instance, at least one edge must be a multiple of 5, one must be a multiple of 7, one must be a multiple of 11, and one must be a multiple of 19. Computer searches have shown that one of the sides must be at least one trillion.

There are some near-misses. The brick with sides 672, 153, and 104 has an integer main diagonal and two of the three lengths for face diagonals are also integers. In 2004 Jorge Sawyer and Clifford Reiter proved that perfect parallelepipeds exist.[91] A parallelepiped (the word comes from parallel-epi-ped, but is often misspelt 'parallelopiped') is like a cuboid but its faces are parallelograms. So it's tilted. The edges have lengths 271, 106, and 103; the minor face diagonals have lengths 101, 266, and 255; the major face diagonals have lengths 183, 312, and 323; and the body diagonals have lengths 374, 300, 278, and 272.

Lonely Runner Conjecture

This comes from an abstruse area of mathematics known as Diophantine approximation theory, and was formulated by Jörg Wills in 1967. Luis Goddyn coined the name in 1998. Suppose that n runners run round a circular track of unit length at uniform speed, with each runner's speed being different. Will every runner be lonely – that is, be more than a distance $1/n$ from all other runners – at some instant of time? Different times for different runners, of course. The conjecture is that the answer is always 'yes', and it has been proved when $n = 4, 5, 6,$ and 7.

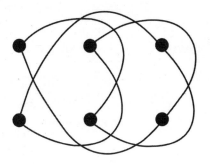

Fig 48 Example of a thrackle.

Conway's Thrackle Conjecture

A thrackle is a network drawn in the plane so that every two edges meet exactly once, Figure 48. They may either meet at a common dot (node, vertex) or they may cross at interior points, but not both. If they cross, they must do so transversely; that is, neither can remain on the same side of the other (which could happen if, for example, they are tangent to each other). In unpublished work, John Horton Conway conjectured that in any thrackle the number of lines is less than or equal to the number of dots. In 2011 Radoslav Fulek and János Pach proved that every thrackle with n dots has at most $1.428n$ lines.[92]

Irrationality of Euler's Constant

There is no known 'closed form' formula for the sum of the harmonic series

$$H_n = 1 + \frac{1}{2} + \frac{1}{3} + \frac{1}{4} + \frac{1}{5} + \cdots + \frac{1}{n}$$

and probably no such formula exists. However, there is an excellent approximation: as n increases, H_n gets ever closer to $\log n + \gamma$. Here γ is Euler's constant, with a numerical value of roughly 0.5772156649. Euler established this formula in 1734, and Lorenzo Mascheroni studied the constant in 1790. Neither used the symbol γ.

Euler's constant is one of those strange numbers that sometimes arise in mathematics, like π and e, which appear all over the place, but seem to be creatures of their own, not expressible in any nice manner in terms of simpler numbers. We saw in chapter 3 that both π and e are transcendental: they do not solve any algebraic equations with integer coefficients. In particular, they are irrational: not exact fractions. It is widely thought that Euler's constant is transcendental, but we don't even know for sure whether it is irrational. If $\gamma = p/q$ for integers p and q, then q is at least $10^{242,080}$.

Euler's constant is important in many areas of mathematics, ranging from the Riemann zeta function to quantum field theory. It appears in many contexts and shows up in many formulas. It is outrageous that we can't decide whether it is rational.

Real Quadratic Number Fields

In chapter 7 we saw that some algebraic number fields have unique prime factorisation and some do not. The best understood algebraic number fields are the quadratic ones, obtained by taking the square root of some number d that is not a perfect square; indeed, has no square prime factors. The corresponding ring of algebraic integers then consists of all number of the form $a + b\sqrt{d}$, where a and b are integers if d is not of the form $4k + 1$, and they are either integers, or are both odd integers divided by 2, if d is of that form.

When d is negative, it is known that prime factorisation is unique for exactly nine values: -1, -2, -3, -7, -11, -19, -43, -67, and -163. Proving uniqueness in these cases is relatively straightforward, but finding whether there are any others is much harder. In 1934 Hans Heilbronn and Edward Linfoot showed that at most one more negative integer can be added to the list. Kurt Heegner gave a proof that the list is complete in 1952, but it was thought to have a gap. In 1967 Harold Stark gave a complete proof, observing that it did not differ significantly from Heegner's – that is, the gap was unimportant. At much the same time, Alan Baker found a different proof.

The case when d is positive is quite different. Factorisation is unique for many more values of d. Up to 50, those values are 2, 3, 5, 6, 7, 11, 13, 14, 17, 19, 21, 22, 23, 29, 31, 33, 37, 38, 41, 43, 46, 47, and computer calculations reveal many more. For all we know, there may be infinitely many positive d for which the corresponding quadratic number field has unique factorisation. A heuristic analysis by Cohen and Lenstra suggests that roughly three quarters of all positive d should define number fields with unique factorisation. Computer results agree with that estimate. The problem is to prove these observations are correct.

Langton's Ant

As the twenty-first century unfolds it has become increasingly apparent that some of the traditional techniques of mathematical modelling are unable to cope with the complexities of the problems facing humanity, such as the global financial system, the dynamics of ecosystems, and the role of genes in the growth of living organisms. Many of these systems involve large numbers of agents – people, companies, organisms, genes – that interact with each other. These interactions can often be modelled quite accurately using simple rules. Over the past 30 years, a new kind of model has appeared, which tries to tackle the behaviour of systems with many agents head-on. To understand how 100,000 people will move round a sports stadium, for example, you don't average them to create a sort of human fluid and ask how it flows. Instead, you build a computer model with 100,000 individual agents, impose suitable rules, and run a simulation to see what this computer crowd does. This kind of model is called a complex system.

To give you a glimpse of this fascinating new area of mathematics, I'm going to describe one of the simplest complex systems, and explain why we don't fully understand it. It is called Langton's ant. Christopher Langton was an early member of the Santa Fe Institute, founded in 1984 by scientists George Cowan, Murray Gell-Mann, and others to promote the theory and applications of complex systems. Langton invented his ant in 1986. Technically it is a cellular automaton, a system of cells in a square grid, whose states are shown by colours. At each time step, the colour of a cell changes according to those of its neighbours.

The rules are absurdly simple. The ant lives on an infinite square grid of cells, initially all of them being white. It carries an inexhaustible pot of quick-drying black paint and another inexhaustible pot of quick-drying white paint. It can face north, east, south, or west; by symmetry we can assume it starts out facing north. At each instant it looks at the colour of the square that it occupies, and changes it from white to black or from black to white using its pots of paint. If its square was white, it then turns 90 degrees to the right and takes one step forward. If its square was black, it then turns 90 degrees to the left and takes one step forward. Now it repeats this behaviour indefinitely.

If you simulate the ant,[93] it starts out by painting simple, fairly symmetric designs of black and white squares. It returns from time to time to a square that it has already visited, but its tour does not close up into a loop because the colour on that square has changed, so it turns the other way on a repeat visit. As the simulation continues, the ant's design becomes chaotic and random. It has no discernible pattern; basically it's just a mess. At that stage you could reasonably imagine that this chaotic behaviour continues indefinitely. After all, when the ant revisits a chaotic region it will make a chaotic series of

Fig 49 Langton's ant's highway.

turns and repaintings. If you carried on with the simulation, the next 10,000 or so steps would appear to justify that conclusion. However, if you keep going, a pattern appears. The ant enters a repeating cycle of 104 steps, at the end of which it has moved two squares diagonally. It then paints a broad diagonal stripe of black and white cells, called a highway, which goes on for ever, Figure 49.

Everything so far described can be proved in full rigour, just by listing the steps that the ant takes. The proof would be quite long – a list of 10,000 steps – but it would still be a proof. But the mathematics gets more interesting if we ask a slightly more general question. Suppose that before the ant starts out, we paint a finite number of squares black. We can choose these squares any way we like: random dots, a solid rectangle, the Mona Lisa. We can use a million of them, or a billion, but not infinitely many. What happens?

The ant's initial excursions change dramatically whenever it meets one of our new black squares. It can potter around all over the place, drawing intricate shapes and redrawing them... But in every simulation yet performed, no matter what the initial configuration might be, the ant eventually settles down to building its highway, using the same 104-step cycle. Does this always happen? Is the highway the unique 'attractor' for ant dynamics? Nobody knows. It is one of the basic unsolved problems of complexity theory. The best we know is that whatever the initial configuration of black cells may be, the ant cannot remain forever inside a bounded region of the grid.

Fig 50 Hadamard matrices of size 2, 4, 8, 12, 16, 20, 24, and 28.
http://mathworld.wolfram.com/HadamardMatrix.html

Hadamard Matrix Conjecture

A Hadamard matrix, named for Jacques Hadamard, is a square array of 0s and 1s such that any two distinct rows or columns agree on half of their entries and disagree on the other half. Using black and white to indicate 1 and 0, Figure 50 shows Hadamard matrices of size 2, 4, 8, 12, 16, 20, 24, and 28. These matrices turn up in many mathematical problems, and in computer science, notably coding theory. (In some applications, among them Hadamard's original motivation, the white squares correspond to −1, not 0.)

Hadamard proved that such matrices exist only when $n = 2$ or n is a multiple of 4. Paley's theorem of 1933 proves that a Hadamard matrix always exists if the size is a multiple of 4 and equal to $2^a(p^b + 1)$ where p is an odd prime. Multiples of 4 not covered by this theorem are 92, 116, 156, 172, 184, 188, 232, 236, 260, 268, and other larger values. The conjecture states that a Hadamard matrix exists whenever the size is a multiple of 4. In 1985 K. Sawade found one of size 268; the other numbers not covered by Paley's theorem had already been dealt with. In 2004 Hadi Kharaghani and Behruz Tayfeh-Rezaie found a Hadamard matrix of size 428, and the smallest size for which the answer is not known is now 668.

Fermat-Catalan Equation

This is the Diophantine equation $x^a + y^b = z^c$ where a, b, and c are positive integers, the exponents. I will call it the Fermat-Catalan equation because its solutions relate both to Fermat's last theorem, chapter 7, and to the Catalan conjecture, chapter 6. If a, b, and c are small, nonzero integer solutions are not especially surprising. For example, if they are all 2, then we have the Pythagorean equation, known since the time of Euclid to have infinitely many solutions. So the main interest is in the cases when these exponents are large. The technical definition of 'large' is that $s = 1/a + 1/b + 1/c$ is less than 1. Only ten large solutions of the Fermat-Catalan equation are known:

$$1 + 2^3 = 3^2 \qquad 17^7 + 76271^3 = 21063928^2$$
$$2^5 + 7^2 = 3^4 \qquad 1414^3 + 2213459^2 = 65^7$$
$$7^3 + 13^2 = 2^9 \qquad 9262^3 + 15312283^2 = 113^7$$
$$2^7 + 17^3 = 71^2 \qquad 43^8 + 9622^3 = 30042907^2$$
$$3^5 + 11^4 = 122^2 \qquad 33^8 + 159034^2 = 15613^3.$$

The first of these is considered large because $1 = 1^a$ for any a, and $a = 7$ satisfies the definition. The Fermat-Catalan conjecture states that the Fermat-Catalan equation has only finitely many integer solutions, without a common factor, when s is large. The main result was proved in 1997 by Henri Darmon and Loïc Merel: there are no solutions in which $c = 3$ and a and b are equal and greater than 3. Little else is known. Further progress seems to depend on a fascinating new conjecture, which comes next.

ABC Conjecture

In 1983 Richard Mason noticed that one case of Fermat's last theorem had been ignored: first powers. That is, consider the equation $a + b = c$.

At first sight this idea is completely pointless. It takes very little grasp of algebra to solve this equation for any of the three variables in terms of the other two. For example $a = c - b$. What changes the whole game, though, is context. Mason realised that everything becomes much deeper if we ask the right questions about a, b, and c. The outcome of his extraordinary idea was a new conjecture in number theory with far-reaching consequences. It could dispose of many currently unsolved problems and lead to better and simpler proofs of some of the biggest theorems in number theory. This is the ABC conjecture, and it is supported by a huge quantity of numerical evidence. It rests on a loose analogy between integers and polynomials.

Euclid and Diophantus knew a recipe for Pythagorean triples, which we now write as a formula, chapter 6. Can this trick be repeated with other equations? In 1851 Joseph Liouville proved that no such formula exists for the Fermat equation when the power is 3 or greater. Mason applied similar reasoning to the simpler equation

$$a(x) + b(x) = c(x)$$

for three polynomials. It's an outrageous idea, because all solutions can be found using elementary algebra. The main result, though, is elegant and far from obvious: if each polynomial has a factor that is a square, a cube, or a higher power, the equation has no solutions.

Theorems about polynomials often have analogues about integers. In particular, irreducible polynomials correspond to prime numbers. The natural analogue for integers of Mason's theorem about polynomials goes as follows. Suppose $a + b = c$ where a, b, and c are integers with no common factor; then the number of prime factors of each of a, b, and c is less than the number of *distinct* prime factors of abc. Unfortunately, simple examples show that this is false. In 1985 David Masser and Joseph Oesterlé modified the statement and proposed a version of this conjecture that did not conflict with any known examples. Their ABC conjecture may well be the biggest open question in number theory at the present time.[94] If someone proved the ABC conjecture tomorrow, many deep and difficult theorems, proved over past decades with enormous insight and effort, would have new, simple proofs. Another consequence would be Marshall Hall's conjecture: the difference between any perfect cube and any perfect square has to be fairly large. Yet another potential application of the ABC conjecture is to Brocard's problem, the first in this chapter. In 1993 Marius Overholt proved that if the ABC conjecture is true, there are only finitely many solutions to Brocard's equation.

One of the most interesting consequences of the ABC conjecture relates to the Mordell conjecture. Faltings has proved this using sophisticated methods, but his result would be even more powerful if we knew one extra piece of information: a bound on the size of the solutions. Then there would exist an algorithm to find them all. In 1991 Noam Elkies showed that a specific version of the ABC conjecture, in which various constants that appear are bounded, implies this improvement on Faltings's theorem. Laurent Moret-Bailly showed that the converse is true, in a very strong way. Sufficiently strong bounds on the size of solutions of *just one* Diophantine equation, $y^2 = x^5 - x$, imply the full ABC conjecture. Although it is not as well known as many other unsolved conjectures, the ABC conjecture is undoubtedly one of the great problems of mathematics. According to Granville and Thomas Tucker, disposing of it would have 'an extraordinary impact on our understanding of number theory. Proving or disproving it would be amazing.'[95]

Glossary

Algebraic integer. A complex number that satisfies a polynomial equation with integer coefficients and highest coefficient 1. For example $i\sqrt{2}$, which satisfies the equation $x^2 + 2 = 0$.

Algebraic number. A complex number that satisfies a polynomial equation with integer coefficients, or equivalently rational coefficients. For example $i\sqrt{2}/3$, which satisfies the equation $x^2 + \frac{2}{9} = 0$, or equivalently $9x^2 + 2 = 0$

Algebraic variety. A multidimensional space defined by a set of algebraic equations.

Algorithm. A specified procedure to solve a problem, guaranteed to stop with an answer.

Angular momentum. A measure of how much spin a body has.

Arithmetic sequence. A sequence of numbers in which each successive number is the previous one plus a fixed amount, the common difference. For example, 2, 5, 8, 11, 14, ... with common difference 3. The older term is 'arithmetic progression'.

Asymptotic. Two quantities defined in terms of a variable are asymptotic if their ratio gets closer and closer to 1 as the variable becomes arbitrarily large.

Axis of rotation. A fixed line about which some object rotates.

Ball. A solid sphere – that is, a sphere and its interior.

Blowup time. The time beyond which a solution of a differential equation fails to exist.

Boundary. The edge of a specified region.

Chaos. Apparently random behaviour in a deterministic system.

Class E. An algorithm whose running time, for an input of size n, resembles the nth power of some constant.

Class P. An algorithm whose running time resembles some fixed power of the input size.

Class not-P. Not class P.

Class NP. A problem for which a proposed solution can be checked (but not necessarily found) by a class P algorithm.

Coefficient. In a polynomial such as $6x^3 - 5x^2 + 4x - 7$, the coefficients are the numbers $6, -5, 4, -7$ that multiply the various powers of x.

Cohomology group. An abstract algebraic structure associated with a topological space, analogous to but 'dual' to the homology group.

Complex analysis. Analysis – logically rigorous calculus – carried out with complex-valued functions of a complex variable.

Complex number. A number of the form $a + bi$ where i is the square root of minus one and a, b are real numbers.

Composite number. A whole number that can be obtained by multiplying together two smaller whole numbers.

Congruent number. A number that can be the common difference of a sequence of three squares of rational numbers.

Continuous transformation. A transformation of a space with the property that points that are very close together do not get pulled a long way apart.

Coordinate. One number in a list that determines the position of a point on a plane or in space.

Cosine. A trigonometric function of an angle, defined by $\cos A = a/c$ in Figure 51.

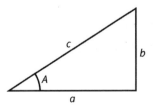

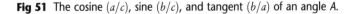

Fig 51 The cosine (a/c), sine (b/c), and tangent (b/a) of an angle A.

Counterexample. An example that disproves some statement. For instance, 9 is a counterexample to the statement 'all odd numbers are prime'.

Cube. A number multiplied by itself and then again by itself. For example, the cube of 7 is $7 \times 7 \times 7 = 343$. Usually written as 7^3.

Cubic equation. Any equation $ax^3 + bx^2 + cx + d = 0$ where x is unknown and a, b, c, d are constants.

Curvature. A measure of how space curves near a given point. A sphere has positive curvature, a plane has zero curvature, and a saddle-shaped space has negative curvature.

Cycle. In topology: a formal combination of loops in a triangulation with numerical labels attached. In algebraic geometry: a formal combination of subvarieties with numerical labels attached.

Cyclotomic integer/number. A sum of powers of a complex root of unity with integer/rational coefficients.

Degree. The highest power of the variable that occurs in a polynomial. For example the degree of $6x^3 - 5x^2 + 4x - 7$ is 3.

Differential equation. An equation relating a function to its rate of change.

Dimension. The number of coordinates required to specify the location of a point in a given space. For example the plane has dimension 2 and the space we live in (as modelled by Euclid's geometry) has dimension 3.

Diophantine equation. An equation for which solutions are required to be rational numbers.

Dirichlet L-function. A generalisation of the Riemann zeta function.

Disc (topological). A region in a surface that can be deformed continuously into a circle plus its interior.

Dodecahedron. A solid whose faces are 12 regular pentagons. See Figure 38.

Dual network. A network obtained from a given network by associating a point with every region, and joining points by edges if the corresponding regions are adjacent. See Figure 10.

Dynamical system. Any system that changes over time according to specified rules. For example, the motion of the planets in the solar system.

Eigenvalue. One of a set of special numbers associated with an operator. If the operator applied to some vector yields a constant multiple of that vector, the multiple concerned is an eigenvalue.

Electromagnetic field. A function that specifies the strengths and directions of the electric and magnetic fields at any point in space.

Elliptic curve. A curve in the plane whose equation has the form $y^2 = ax^3 + bx^2 + cx + d$ for constants a, b, c, d, usually assumed to be rational. See Figure 27.

Elliptic function. A complex function that remains unchanged when two independent complex numbers are added to its variable. That is, $f(z) = f(z + u) = f(z + v)$ where v is not a real multiple of u. See Figure 30.

Euler characteristic. $F - E + V$ where F is the number of faces in a triangulation of some space, E is the number of edges, and V is the number of vertexes. For a torus with g holes it equals $2 - 2g$, whatever the triangulation may be.

Euler's constant. A special number denoted by γ, approximately equal to 0.57721. See Note 67.

Exponent. In a power of a variable x, the exponent is the power concerned. For example, in x^7 the exponent is 7.

Face-centred cubic lattice. A repeating set of points in space, obtained by stacking cubes together like a three-dimensional chessboard, and then taking the corners of the cubes and the centres of their six square faces. See Figures 17, 19.

Factorisation. The process that writes a number in terms of its prime divisors. For example, the factorisation of 60 into primes is $2^2 \times 3 \times 5$.

Fermat number. A number of the form $2^{2^k} + 1$ where k is a whole number. If this number is prime then it is called a *Fermat prime*.

Flat torus. A torus obtained by identifying opposite edges of a square, whose natural geometry has zero curvature. See Figure 12.

Function. A rule f which, when applied to a number x, produces another number $f(x)$. For example, if $f(x) = \log x$ then f is the logarithm function. The variable x can be real or complex (in which case it is often written as z). More generally, x and $f(x)$ can be members of specified sets; in particular, the plane or space.

Fundamental group. The group formed by homotopy classes of loops in some topological space, under the operation 'travel along the first loop and then along the second'.

Gauge symmetry. A group of local symmetries of a system of equations: transformations of the variables that can vary from point to point in space, with the property that any solution of the equations remains a solution provided a compensating change with a sensible physical interpretation is made to the equations.

Gauge theory. A quantum field theory with a group of gauge symmetries.

General relativity. Einstein's theory of gravitation, in which the force of gravity is interpreted as the curvature of space-time.

Genus. The number of holes in a surface.

Group. An abstract algebraic structure comprising a set and a rule for combining any two elements of the set, subject to three conditions: the associative law, the existence of an identity element, and the existence of inverses.

Higgs boson. A fundamental particle whose existence explains why all particles have masses. Its discovery by the Large Hadron Collider was announced in July 2012.

Hodge class. A cohomology class of cycles on an algebraic variety with special analytic properties.

Homology (group). A topological invariant of a space, defined by closed loops. Two such loops are homologous if their difference is the boundary of a topological disc.

Homotopy (group). A topological invariant of a space, defined by closed loops. Two such loops are homotopic if each can be continuously deformed into the other.

Ideal (number). A number that is not contained in a given system of algebraic numbers, but is related to that system in a way that restores unique prime factorisation in cases when that property fails. Replaced in modern algebra by an ideal, which is a special kind of subset of the system concerned.

Induction. A general method for proving theorems about whole numbers. If some property is valid for 0, and its validity for any whole number n implies its validity for $n + 1$, then the property is valid for all whole numbers.

Integer. Any of the numbers $\ldots, -3, -2, -1, 0, 1, 2, 3, \ldots$.

Integral. An operation of the calculus, which in effect adds together large numbers of small contributions. The integral of a function is the area under its graph.

Irrational number. A real number that is not rational; that is, not of the form p/q where p and q are integers and $q \neq 0$. Examples are $\sqrt{2}$ and π.

Irreducible polynomial. A polynomial that cannot be obtained by multiplying two polynomials of smaller degree.

Lattice. In the plane: a set of points that repeats its form along two independent directions, like wallpaper patterns, see Figure 26. In space: a set of points that repeats its form along three independent directions, like the atoms in a crystal.

Lattice packing. A collection of identical circles or spheres whose centres form a lattice.

Logarithm. The (natural) logarithm of x, written $\log x$, is the power to which e $(= 2.71828\ldots)$ must be raised to obtain x. That is, $e^{\log x} = x$.

Logarithmic integral. The function $\text{Li}(x) = \int_0^x \frac{dt}{\log t}$.

Loop. A closed curve in a topological space.

Manifold. A multidimensional analogue of a smooth surface.

Maximum. The largest value of something.

Minimal criminal. A mathematical object that does not possess some desired property, and in some sense is the smallest possible such object. For example, a map that cannot be coloured with four colours, and also has the smallest number of regions for which this can occur. Minimal criminals are often hypothetical, and the aim is to prove that they don't exist.

Minimum. The smallest value of something.

Modular arithmetic. A system of arithmetic in which multiples of some specific number, called the *modulus*, are treated as if they are all zero.

Momentum. Mass multiplied by velocity.

Network. A set of points (nodes, dots) joined by lines (edges).

Non-Euclidean geometry. An alternative to Euclid's geometry in which all of the usual properties of points and lines remain valid, except for the existence of a unique line parallel to a given line and passing through a given point. There are two kinds: elliptic and hyperbolic.

NP-complete. A specific class NP problem, with the property that if there exists a class P algorithm to solve it, then *any* NP problem can be solved using a class P algorithm.

Operator. A special kind of function A, which when applied to a vector v yields another vector Av. It must satisfy the linearity conditions $A(v + w) = Av + Aw$ and $A(av) = aA(v)$ for any constant a.

Optimisation. Finding the maximum or minimum of some function.

Packing. A collection of shapes arranged in space so that they do not overlap.

Partial differential equation. A differential equation involving the rates of change of some function with respect to two or more different variables (often space and time).

Particle. A mass concentrated at one point.

Pentagon. A polygon with five sides.

Periodic. Anything that repeats the same behaviour indefinitely.

Phase. A complex number on the unit circle used to multiply a quantum wavefunction.

Polygon. A flat shape whose boundary consists of a finite number of straight lines.

Polyhedron. A solid whose boundary consists of a finite number of polygons.

Polynomial. An algebraic expression like $6x^3 - 5x^2 + 4x - 7$, in which powers of a variable x are multiplied by constants and added together.

Power. A number multiplied by itself a specified number of times. For example, the fourth power of 3 is $3 \times 3 \times 3 \times 3 = 81$, symbolised as 3^4.

Power series. Like a polynomial except that infinitely many powers of the variable can occur – for example $1 + 2x + 3x^2 + 4x^3 + \cdots$. In suitable circumstances this infinite sum can be assigned a well-defined value, and the series is said to converge.

Prime ideal. An analogue of a prime number for algebraic number systems.

Prime number. A whole number greater than 1 that cannot be obtained by multiplying two smaller whole numbers. The first few prime numbers are 2, 3, 5, 7, 11, 13.

Projective geometry. A type of geometry in which parallel lines do not exist: any two lines meet at a single point. Obtained from Euclidean geometry by adding a new 'line at infinity'.

Pythagorean triple. Three whole numbers a, b, c such that $a^2 + b^2 = c^2$. For example, $a = 3, b = 4, c = 5$. By Pythagoras's theorem, numbers of this type form the sides of a right-angled triangle.

Quadratic equation. Any equation $ax^2 + bx + c = 0$ where x is unknown and a, b, c are constants.

Quantum field theory. A quantum-mechanical theory of a quantity that pervades space and can (and usually does) have different values at different locations.

Quantum wavefunction. A mathematical function determining the properties of a quantum system.

Rank. The largest number of independent rational solutions of the equation defining an elliptic curve. 'Independent' means that they cannot be deduced from the other solutions using a standard geometric construction that combines any two solutions to yield a third, see Figure 25.

Ratio. The ratio of two numbers a and b is a/b.

Rational number. A real number of the form p/q where p and q are integers and $q \neq 0$. An example is $22/7$.

Real number. Any number that can be expressed in decimals, possibly going on for ever – for example, $\pi = 3.1415926535897932385\ldots$.

Reducible configuration. A part of a network with the following property: if the network obtained by removing it can be coloured with four colours, so can the original network.

Regular polygon. A polygon whose sides all have the same length, and whose angles are all equal. See Figure 4.

Regular solid. A solid whose boundary is composed of identical regular polygons, arranged in the same manner at every corner. Euclid proved that exactly five regular solids exist.

Rhombic dodecahedron. A solid whose boundary is composed of 12 identical rhombuses – parallelograms with all sides equal. See Figure 15.

Ricci flow. An equation prescribing how the curvature of a space changes over time.

Root of unity. A complex number ζ for which some power ζ^k is 1. See Figure 7 and Note 53.

Rotation. In the plane: a transformation in which all points move through the same angle about a fixed centre. In space: a transformation in which all points move through the same angle about a fixed line, the axis.

Ruler-and-compass construction. Any geometric construction that can be performed using an unmarked ruler and a compass (strictly: a pair of compasses).

Sequence. A list of numbers arranged in order. For example, the sequence 1, 2, 4, 8, 16, ... of powers of 2.

Series. An expression in which many quantities – often infinitely many – are added together.

Set. A collection of (mathematical) objects. For example, the set of all whole numbers.

Sine. A trigonometric function of an angle, defined by $\sin A = b/c$ in Figure 51.

Singularity. A point at which something nasty happens, such as a function becoming infinite or a solution of some equation failing to exist.

Sphere. The set of all points in space at a given distance from some fixed point, the centre. It is round, like a ball, but the term 'sphere' refers only to the points on the surface of the ball, not inside it.

3-Sphere. Three-dimensional analogue of a sphere: the set of all points in four-dimensional space at a given distance from some fixed point, the centre.

Square. A number multiplied by itself. For example, the square of 7 is $7 \times 7 = 49$, symbolised as 7^2.

Stable. A state of a dynamical system to which it returns if it is subjected to a small disturbance.

Standard Model. A quantum-mechanical model that accounts for all known fundamental particles.

Surface. A shape in space obtained by patching together regions that are topologically equivalent to the inside of a circle. Examples are the sphere and torus.

Symmetry. A transformation of some object that leaves its overall form unchanged. For example, rotating a square through a right angle.

Tangent. A trigonometric function of an angle, defined by $\tan A = b/a$ in Figure 51.

Topological space. A shape that is considered to be 'the same' if it is subjected to any continuous transformation.

Topology. The study of topological spaces.

Torus. A surface like that of a doughnut with one hole. See Figure 12.

Transcendental number. A number that does not satisfy any algebraic equation with rational coefficients. Examples are π and e.

Transformation. Another word for 'function', commonly used when the variables involved are points in some space. For example, 'rotate about the centre through a right angle' is a transformation of a square.

Translation. A transformation of space in which all points slide through the same distance and in the same direction.

Triangulation. Splitting a surface into a network of triangles, or its multidimensional analogue.

Trisection. Dividing into three equal parts, especially in connection with angles.

Trivial group. A group consisting only of a single element, the identity.

Unavoidable configuration. A member of a list of networks, at least one of which must occur in any network in the plane.

Unique prime factorisation. The property that any number can be written as a product of prime numbers in only one way, except for changing the order in which the factors are written. This property is valid for integers, but can fail in more general algebraic systems.

Unstable. A state of a dynamical system to which it may not return if it is subjected to a small disturbance.

Upper bound. A specific number that is guaranteed to be bigger than some quantity whose size is being sought.

Variable. A quantity that can take on any value in some range.

Variety. A shape in space defined by a system of polynomial equations.

Vector. In mechanics, a quantity with both size and direction. In algebra and analysis, a generalisation of this idea.

Velocity. The rate at which position changes with respect to time. Velocity has both a size (called speed) and a direction.

Velocity field. A function that specifies a velocity at each point in space. For example, when a fluid flows, its velocity can be specified at each point, and typically differs at different points.

Vortex. Fluid flowing round and round like a whirlpool. May be any size, including very small.

Wave. A disturbance that moves through a medium such as a solid, liquid, or gas without making any permanent change to the medium.

Whole number. Any of the numbers 0, 1, 2, 3, ...

Winding number. The number of times that a curve winds anticlockwise round some chosen point.

Zero (of a function). If f is a function, then x is a zero of f if $f(x) = 0$.

Zeta function. A complex function introduced by Riemann that represents the prime numbers analytically. It is defined by the series

$$\zeta(s) = \frac{1}{1^s} + \frac{1}{2^s} + \frac{1}{3^s} + \frac{1}{4^s} + \frac{1}{5^s} + \frac{1}{6^s} + \frac{1}{7^s} + \cdots$$

which converges when the real part of s is greater than 1. This definition can be extended to all complex s, except 1, by a process called analytic continuation.

Further reading

*Books marked * are technical.*

* Colin C. Adams, *The Knot Book*, W.H. Freeman, 1994.

* Felix Browder (ed.), *Mathematical Developments Arising from Hilbert Problems* (2 vols), Proceedings of Symposia in Pure Mathematics 28, American Mathematical Society, 1976.

* Tian Yu Cao, *Conceptual Developments of 20th Century Field Theories*, Cambridge University Press, 1997.

William J. Cook, *In Pursuit of the Travelling Salesman*, Princeton University Press, 2012.

Keith Devlin, *The Millennium Problems*, Granta, 2004.

Florin Diacu and Philip Holmes, *Celestial Encounters*, Princeton University Press, 1999.

Underwood Dudley, *A Budget of Trisections*, Springer, 1987.

Underwood Dudley, *Mathematical Cranks*, Mathematical Association of America, 1992.

Marcus Du Sautoy, *The Music of the Primes*, Harper Perennial, 2004.

Masha Gessen, *Perfect Rigour*, Houghton Mifflin, 2009.

* Jay R. Goldman, *The Queen of Mathematics*, A.K. Peters, 1998.

Jacques Hadamard, *The Psychology of Invention in the Mathematical Field*, Dover, 1954.

* Harris Hancock, *Lectures on the Theory of Elliptic Functions*, Dover, 1958.

Michio Kaku, *Hyperspace*, Oxford University Press, 1994.

* Jeffrey C. Lagarias, *The Ultimate Challenge: The 3x+1 Problem*, American Mathematical Society, 2011.

* Charles Livingston, *Knot Theory*, Carus Mathematical Monographs **24**, Mathematical Association of America, 1993.

Mario Livio, *The Equation That Couldn't Be Solved*, Simon and Schuster, 2005.

* Henry McKean and Victor Moll, *Elliptic Curves*, Cambridge University Press, 1997.

Donal O'Shea, *The Poincaré Conjecture*, Walker, 2007.

Lisa Randall, *Warped Passages*, Allen Lane, 2005.

* Gerhard Ringel, *Map Color Theorem*, Springer, 1974.

* C. Ambrose Rogers, *Packing and Covering*, Cambridge Tracts in Mathematics and Mathematical Physics **54**, Cambridge University Press, 1964.

Karl Sabbagh, *Dr Riemann's Zeros*, Atlantic Books, 2002.

Ian Sample, *Massive*, Basic Books, 2010.

* René Schoof, *Catalan's Conjecture*, Springer, 2008.

Simon Singh, *Fermat's Last Theorem*, Fourth Estate, 1997.

Ian Stewart, *From Here to Infinity*, Oxford University Press, 1996.

Ian Stewart, *Why Beauty is Truth*, Basic Books, 2007.

Ian Stewart, *Seventeen Equations that Changed the World*, Profile, 2012.

George Szpiro, *Kepler's Conjecture*, Wiley, 2003.

* Jean-Pierre Tignol, *Galois' Theory of Algebraic Equations*, Longman Scientific and Technical, 1980.

Matthew Watkins, The Mystery of the Prime Numbers, *Inamorata Press, 2010.*

Robin Wilson, *Four Colours Suffice*, Allen Lane, 2002.

Benjamin Yandell, *The Honors Class*, A.K. Peters, 2002.

Notes

1 The German original is: 'Wir müssen wissen. Wir werden wissen.' It occurs in a speech that Hilbert recorded for radio. See Constance Reid, *Hilbert*, Springer, Berlin, 1970, page 196.

2 Simon Singh, *Fermat's Last Theorem*, Fourth Estate, 1997.

3 Gauss, letter to Heinrich Olbers, 21 March 1816.

4 Wiles's title was 'Modular curves, elliptic forms, and Galois representations'.

5 Andrew Wiles, Modular elliptic curves and Fermat's last theorem, *Annals of Mathematics* **141** (1995) 443–551.

6 Ian Stewart, *Seventeen Equations that Changed the World*, Profile 2012, chapter 11.

7 Ibid., chapter 9.

8 Hilbert's problems, and their current status, edited slightly from *Professor Stewart's Hoard of Mathematical Treasures*, Profile 2009, are as follows:

1 **Continuum Hypothesis:** Is there an infinite cardinal number strictly between the cardinalities of the integers and the real numbers? Solved by Paul Cohen in 1963 – the answer depends on which axioms you use for set theory.

2 **Logical Consistency of Arithmetic:** Prove that the standard axioms of arithmetic can never lead to a contradiction. Solved by Kurt Gödel in 1931: impossible with the usual axioms for set theory.

3 **Equality of Volumes of Tetrahedrons:** If two tetrahedrons have the same volume, can you always cut one into finitely many polygonal pieces, and reassemble them to form the other? Solved in 1901 by Max Dehn, in the negative.

4 **Straight Line as Shortest Distance between Two Points:** Formulate axioms for geometry in terms of the above definition of 'straight line', and investigate the implications. Too broad to have a definitive solution, but much work has been done.

5 **Lie Groups without Assuming Differentiability:** Technical issue in the theory of groups of transformations. In one interpretation, solved by Andrew Gleason in the 1950s. In another, by Hidehiko Yamabe.

6 **Axioms for Physics:** Develop a rigorous system of axioms for mathematical areas of physics, such as probability and mechanics. Andrei Kolmogorov axiomatised probability in 1933.

7 **Irrational and Transcendental Numbers:** Prove that certain numbers are irrational or transcendental. Solved by Aleksandr Gelfond and Theodor Schneider in 1934.

8 **Riemann Hypothesis:** Prove that all nontrivial zeros of Riemann's zeta-function lie on the critical line. See chapter 9.

9 **Laws of Reciprocity in Number Fields:** Generalise the classical law of quadratic reciprocity, about squares to some modulus, to higher powers. Partially solved.

10 **Determine When a Diophantine Equation Has Solutions:** Find an algorithm which, when presented with a polynomial equation in many variables, determines whether any solutions in whole numbers exist. Proved impossible by Yuri Matiyasevich in 1970.

11 **Quadratic Forms with Algebraic Numbers as Coefficients:** Technical issues about the solution of many-variable Diophantine equations. Partially solved.

12 **Kronecker's Theorem on Abelian Fields:** Technical issues generalising a theorem of Kronecker. Still unsolved.

13 **Solving Seventh-Degree Equations using Special Functions:** Prove that the general seventh-degree equation cannot be solved using functions of two variables. One interpretation disproved by Andrei Kolmogorov and Vladimir Arnold.

14 **Finiteness of Complete Systems of Functions:** Extend a theorem of Hilbert about algebraic invariants to all transformation groups. Proved false by Masayoshi Nagata in 1959.

15 **Schubert's Enumerative Calculus:** Hermann Schubert found a non-rigorous method for counting various geometric configurations. Make the method rigorous. No complete solution yet.

16 **Topology of Curves and Surfaces:** How many connected components can an algebraic curve of given degree have? How many distinct periodic cycles can an algebraic differential equation of given degree have? Limited progress.

17 **Expressing Definite Forms by Squares:** If a rational function always takes non-negative values, must it be a sum of squares? Solved by Emil Artin, D.W. Dubois, and Albrecht Pfister. True over the real numbers, false in some other number systems.

18 Tiling Space with Polyhedrons: General issues about filling space with congruent polyhedrons. Also mentions the Kepler conjecture, now proved, see chapter 5.

19 Analyticity of Solutions in Calculus of Variations: The calculus of variations answers questions like: 'Find the shortest curve with the following properties.' If such a problem is defined by nice functions, must the solution also be nice? Proved by Ennio de Giorgi in 1957, and by John Nash.

20 Boundary Value Problems: Understand the solutions of the differential equations of physics, inside some region of space, when properties of the solution on the boundary of that region are prescribed. Essentially solved, by numerous mathematicians.

21 Existence of Differential Equations with Given Monodromy: A special type of complex differential equation can be understood in terms of its singular points and its monodromy group. Prove that any combination of these data can occur. Answered yes or no, depending on interpretation.

22 Uniformisation using Automorphic Functions: Technical question about simplifying equations. Solved by Paul Koebe soon after 1900.

23 Development of Calculus of Variations: Hilbert appealed for fresh ideas in the calculus of variations. Much work done; question too vague to be considered solved.

9   Reprinted as: Jacques Hadamard, *The Psychology of Invention in the Mathematical Field*, Dover, 1954.

10   The Agrawal-Kayal-Saxena algorithm is as follows:

Input: integer n.

1   If n is an exact power of any smaller number, output COMPOSITE and stop.

2   Find the smallest r such that the smallest power of r that equals 1 to the modulus n is at least $(\log n)^2$.

3   If any number less than or equal to r has a factor in common with n, output COMPOSITE and stop.

4   If n is less than or equal to r, output PRIME and stop.

5   For all whole numbers a ranging from 1 to a specified limit, check whether the polynomial $(x + a)^n$ is the same as $x^n + a$, to the modulus n and to the modulus $x^r - 1$. If equality holds in any case, output COMPOSITE and stop.

6   Output PRIME.

11 An example of what I have in mind is the formula $[A^{3^n}]$, where the

brackets denote the largest integer less than or equal to their contents. In 1947 W.H. Mills proved that there exists a real constant A such that this formula is prime for any n. Assuming the Riemann hypothesis, the smallest value of A that works is about 1.306. However, the constant is defined using a suitable sequence of primes, and the formula is just a symbolic way to reproduce this sequence. For more such formulas, including some that represent all primes, see

http://mathworld.wolfram.com/PrimeFormulas.html

http://en.wikipedia.org/wiki/Formula_for_primes

12 If n is odd then $n - 3$ is even, and if n is greater than 5 then $n - 3$ is greater than 2. By the first conjecture, $n - 3 = p + q$, so $n = p + q + 3$.

13 I prefer this term to the old-fashioned, but perhaps more familiar, 'arithmetic progression'. No one talks of progressions any more, except for arithmetic and geometric ones. Time to move on.

14 http://www.numberworld.org/misc_runs/pi-5t/details.html

15 My pet hate in this context is 'quantum leap'. In colloquial parlance it indicates some gigantic step forward, or some huge change, like the European discovery of America. In quantum theory, however, a quantum leap is so tiny that no known instrument can observe it directly, a change whose size is about 0.000 ... 01 with 40 or so zeros.

16 Finding a finite dissection of a square into a circle is called Tarski's circle-squaring problem. Miklós Laczkovich solved it in 1990. His method is non-constructive and makes use of the axiom of choice. The number of pieces required is huge, about 10^{50}.

17 The bizarre claims of circle-squarers and angle-trisectors are explored in depth in Underwood Dudley, *A Budget of Trisections*, Springer, 1987, and *Mathematical Cranks*, Mathematical Association of America, 1992. The phenomenon is not new: see Augustus De Morgan, *A Budget of Paradoxes*, Longmans, 1872; reprinted by Books For Libraries Press, 1915.

18 The quadratrix of Hippias is the curve traced by a vertical line that moves steadily across a rectangle and a line that rotates steadily about the midpoint of the bottom of the rectangle, Figure 52. This relationship turns any question about angle-division into the

corresponding one about line-division. For example, to trisect an angle you just trisect the corresponding line. See http://www.geom.uiuc.edu/~huberty/math5337/groupe/quadratrix.html

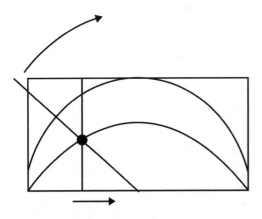

Fig 52 The quadratrix of Hippias (lower curve).

19 Here's an explicit example. Geometrically, if a line meets a circle and is not tangent to it, then it cuts the circle at exactly two points. Consider a line that is parallel to the horizontal axis, distance $\frac{1}{2}$ above it, Figure 53. The equation of this line is very simple: $y = \frac{1}{2}$. (Whatever value x may be, we always get the same value for y.) When $y = \frac{1}{2}$, the equation $x^2 + y^2 = 1$ becomes $x^2 + \frac{1}{4} = 1$. Therefore $x^2 = \frac{3}{4}$, so $x = \frac{\sqrt{3}}{2}$ or $-\frac{\sqrt{3}}{2}$. So algebra tells us that the unit circle meets our chosen line at exactly two points, whose coordinates are $\left(\frac{\sqrt{3}}{2}, \frac{1}{2}\right)$ and $-\left(\frac{\sqrt{3}}{2}, \frac{1}{2}\right)$. This is consistent with Figure 53 and with purely geometrical reasoning.

20 Strictly speaking, the polynomial concerned must have integer coefficients and be irreducible: not the product of two polynomials of lower degree with integer coefficients. Having degree that is a power of 2 is not always sufficient for a ruler-and-compass construction to exist, but it is always necessary. If the degree is not a power of 2, no construction can exist. If it is a power of 2, further analysis is needed to decide whether there is a construction.

21 The converse is also true: given constructions for regular 3- and 5-gons, you can derive one for a 15-gon. The underlying idea is that $2/5 - 1/3 = 1/15$. One subtle point concerns prime powers. The argument doesn't provide a construction for, say, a 9-gon, given one

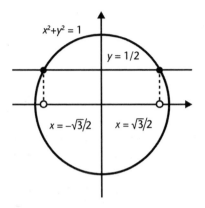

Fig 53 A horizontal line cutting the circle at two points.

for its prime factors – namely, a 3-gon. Gauss proved that no construction is possible for odd prime powers greater than the first.

22 See Ian Stewart, *Seventeen Equations that Changed the World*, Profile, 2012, chapter 5.

23 To make sense of this statement, resolve the quadratic into linear factors. Then $x^2 - 1 = (x + 1)(x - 1)$ which is zero if either factor is zero, so $x = 1$ or -1. The same reasoning can be applied to $x^2 = xx$: this is zero if either the first factor $x = 0$ or the second $x = 0$. It so happens that these two solutions yield the same x, but the occurrence of two factors x distinguishes this situation from something like $x(x - 1)$ where there is only one factor x. When counting how many solutions an algebraic equation has, the answer is generally much tidier if these 'multiplicities' are accounted for.

24 When $n = 9$, the second factor is

$$x^8 + x^7 + x^6 + x^5 + x^4 + x^3 + x^2 + x + 1$$

But this itself has factors: it is equal to

$$\left(x^2 + x + 1\right)\left(x^6 + x^3 + 1\right)$$

Gauss's characterisation of constructible numbers requires each irreducible factor to have degree that is a power of 2. But the second factor has degree 6, which is not a power of 2.

25 Gauss proved that the 17-gon can be constructed provided you can

construct a line whose length is

$$\frac{1}{16}\left[-1+\sqrt{17}+\sqrt{34-2\sqrt{17}}+\right.$$

$$\left.\sqrt{68+12\sqrt{17}-16\sqrt{34+2\sqrt{17}}-2(1-\sqrt{17})(\sqrt{34-2\sqrt{17}})}\right]$$

Since you can always construct square roots, this effectively solves the problem. Other mathematicians found explicit constructions. Ulrich von Huguenin published the first in 1803, and H.W. Richmond found a simpler one in 1893. In Figure 54, take two perpendicular radii AP_0 and BOC of a circle. Make $OJ = 1/4OB$ and angle $OJE = 1/4OJP_0$. Find F so that angle EJF is 45 degrees. Draw a circle with FP_0 as diameter, meeting OB at K. Draw the circle centre E through K, cutting AP_0 in G and H. Draw HP_3 and GP_5 perpendicular to AP_0. Then P_0, P_3, P_5 are respectively the 0th, 3rd, and 5th vertexes of a regular 17-gon, and the other vertexes are now easily constructed.

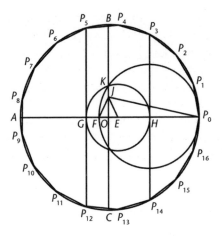

Fig 54 How to construct a regular 17-gon.

26 For the latest discoveries, see Wilfrid Keller, Prime factors of Fermat numbers and complete factoring status:
http://www.prothsearch.net/fermat.html

27 F.J. Richelot published a construction for the regular 257-gon in 1832. J. Hermes of Lingen University devoted ten years to the

65537-gon. His unpublished work can be found at the University of Göttingen, but it is thought to contain errors.

28 A typical continued fraction looks like this:

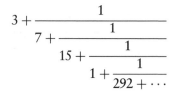

This particular continued fraction is the start of the one that represents π.

29 http://bellard.org/pi-challenge/announce220997.html

30 Louis H. Kauffman, Map coloring and the vector cross product, *Journal of Combinatorial Theory* B **48** (1990) 145–154.

Louis H. Kauffman, Reformulating the map color theorem, *Discrete Mathematics* **302** (2005) 145–172.

31 If the boundaries are allowed to be very complicated indeed, not like a map but far more wiggly, then as many countries as you like can share a common 'border'. A construction called the Lakes of Wada proves this counterintuitive result. See http://en.wikipedia.org/wiki/Lakes_of_Wada

32 The technical term is 'dual graph', because traditionally 'graph' was used instead of 'network'. But 'network' is becoming common, is more evocative, and avoids confusion with other uses of the word 'graph'.

33 Until recently, the *Nature* piece was thought to be the last reference in print to the problem for nearly a century, but mathematical historian Robin Wilson tracked down this subsequent article by Cayley.

34 Working in the dual network, let F be the number of faces (including one big face surrounding the entire network), E the number of edges, and V the number of vertexes. We may assume that every face in the dual network has at least three edges – if it has a face with only two edges then it corresponds to a 'superfluous' vertex of the original network that meets only two edges. This vertex can be deleted and the two edges joined.

Each edge borders two faces, and each face has at least three edges, so $E \geqslant 3F/2$, or equivalently $2E/3 \geqslant F$. By Euler's theorem $F + V - E = 2$, so $2E/3 + V - E \geqslant 2$, which implies that

$$12 + 2E \leqslant 6V$$

Suppose that V_m is the number of vertexes with m neighbours. Then V_2, V_3, V_4, and V_5 are zero. Therefore

$$V = V_6 + V_7 + V_8 + \cdots$$

Since every edge joins two vertexes,

$$2E = 6V_6 + 7V_7 + 8V_8 + \cdots$$

Substituting these into the inequality we get

$$12 + 6V_6 + 7V_7 + 8V_8 + \cdots \leqslant 6V_6 + 6V_7 + 6V_8 + \cdots$$

so that

$$12 + V_7 + 2V_8 + \cdots \leqslant 0$$

which is impossible.

35 'Chain' is misleading, since it suggests a linear sequence. A Kempe chain can contain loops, and it can branch.

36 The proof is given in full in Gerhard Ringel, *Map Color Theorem*, Springer, 1974. It divides into 12 cases, depending on whether the genus is of the form $12k, 12k + 1, ..., 12k + 11$. Call these cases 0–11. With finitely many exceptions, the cases were solved as follows:

Case 5: Ringel, 1954.

Cases 3, 7, and 10: Ringel in 1961.

Cases 0 and 4: C.M. Terry, Lloyd Welch, and Youngs in 1963.

Case 1: W. Gustin and Youngs in 1964.

Case 9: Gustin in 1965.

Case 6: Youngs in 1966.

Cases 2, 8, and 11: Ringel and Youngs in 1967.

The exceptions were genus 18, 20, 23 (solved by Yves Mayer in 1967) and 30, 35, 47, 659 (solved by Ringel and Youngs in 1968). They also dealt with the analogous problem for one-sided surfaces (like the Möbius band but lacking edges), which Heawood had also addressed.

37 The remarkable story of how the bug was discovered, and what happened when it was, can be found at http://en.wikipedia.org/wiki/Pentium_FDIV_bug

38 An excellent site for information about the physics of snowflakes is http://www.its.caltech.edu/~atomic/snowcrystals/

39 C.A. Rogers, The packing of equal spheres, *Proceedings of the London Mathematical Society* **8** (1958) 609–620.

40 Since space is infinite, there are infinitely many spheres, so both the space and the spheres have infinite total volume. We can't define the density to be ∞/∞, because that doesn't have a well-defined numerical value. Instead, we consider larger and larger regions of space and take the limiting value of the proportion of these regions that the spheres fill.

41 http://hydra.nat.uni-magdeburg.de/packing/csq/csq49.html

42 C. Song, P. Wang, and H.A. Makse, A phase diagram for jammed matter, *Nature* **453** (29 May 2008) 629–632.

43 Hai-Chau Chang and Lih-Chung Wang, A simple proof of Thue's theorem on circle packing, arXiv:1009.4322v1 (2010).

44 J.H. Lindsey, Sphere packing in $\Re^3$, *Mathematika* **33** (1986) 137–147.

   D.J. Muder, Putting the best face on a Voronoi polyhedron, *Proceedings of the London Mathematical Society* **56** (1988) 329–348.

45 Hales used several different notions for what I am calling a cage. The final one is 'decomposition star'. My description omits some crucial distinctions in order to make the basic idea comprehensible.

46 Suppose the region is a polygon, as in Figure 55. Given any point that is not on the polygon, there exists some straight line from that point that gets outside a big circle containing the polygon, and does not pass through any vertex of the polygon. (There are finitely many vertexes but infinitely many lines to choose from.) This line cuts the polygon a finite number of times, and this number is either odd or even. Define the inside to consist of all points for which the number is odd, and the outside to consist of all points for which it

is even. It is then straightforward to prove that each of these regions is connected, and the polygon separates them.

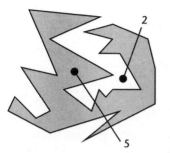

Fig 55 Proving the Jordan curve theorem for a polygon. An odd number of intersections occurs for points in the shaded region (inside), and an even number of intersections occurs for points in the white region (outside).

47 http://code.google.com/p/flyspeck/

48 Andrew Granville and Thomas Tucker, It's as easy as *abc*, *Notices of the American Mathematical Society* **49** (2002) 1224–1231.

49 To expand on this cryptic remark: the formula is

$$\int \frac{dx}{\sqrt{1 - x^2}} = \arcsin x$$

where arcsin (often written $\sin^{-1}$) is the inverse function to the sine. That is, if $y = \sin x$ then $x = \arcsin y$.

50 For example, let k be any complex number, and consider the integral

$$\int \frac{dx}{\sqrt{(1 - x^2)(1 - k^2 x^2)}}$$

This is the inverse function of an elliptic function denoted by sn. There is one such function for each value of k. The set-up is like Note 49, but more elaborate.

51 See Ian Stewart, *Seventeen Equations that Changed the World*, Profile, 2012, chapter 8.

52 The proof can be found in many number theory texts, for example

Gareth A. Jones and J. Mary Jones, *Elementary Number Theory*, Springer, 1998, page 227. On the web, see http://en.wikipedia.org/wiki/Infinite_descent#Non-solvability_of_ r2_.2B_s4_.3D_t4

53 One pth root of unity is the complex number

$$\zeta = \cos 2\pi/p + i \sin 2\pi/p$$

and the others are its powers $\zeta^2, \zeta^3, \ldots \zeta^{p-1}$. To see why, recall that the trigonometric functions sine and cosine are defined using a right-angled triangle, Figure 56 (left). For the angle A, using the traditional a, b, c for the three sides, we define the sine (sin) and cosine (cos) of A by

$$\sin A = a/c \cos A = b/c$$

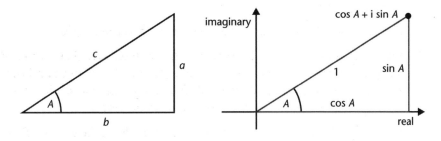

Fig 56 *Left:* Defining the sine and cosine. *Right:* Interpretation in the complex plane.

If we let $c = 1$ and place the triangle in the complex plane, as in Figure 56 (right), the vertex at which c and a meet is the point

$$\cos A + i\sin A$$

It is now straightforward to prove that for any angles A and B,

$$(\cos A + i \sin A)(\cos B + i \sin A) = \cos(A + B) + i \sin(A + B)$$

and this leads directly to De Moivre's formula

$$(\cos A + i \sin A)^n = (\cos nA + i \sin nA)$$

for any positive integer n. Therefore

$$\zeta^p = (\cos 2\pi/p + i \sin 2\pi/p)^p = \cos 2\pi + i \sin 2\pi = 1$$

so each power $1, \zeta, \zeta^2, \zeta^3, ..., \zeta^{p-1}$ is a pth a root of unity. We stop there because $\zeta^p = 1$, so no new numbers arise if we take higher powers.

54 Introduce the *norm*

$$N(a + b\sqrt{15}) = a^2 - 15b^2$$

which has the lovely property

$$N(xy) = N(x)N(y)$$

Then

$$N(2) = 4 \, N(5) = 25 \, N(5 + \sqrt{15}) = 10 \, N(5 - \sqrt{15}) = 10$$

Any proper divisor of one of these four numbers must have norm either 2 or 5 (the proper divisors of their norms). But the equations $a^2 - 15b^2 = 2$ and $a^2 - 15b^2 = 5$ have no integer solutions. Therefore no proper divisors exist.

55 Simon Singh, *Fermat's Last Theorem*, Fourth Estate, 1997.

56 Or maybe not. Vladimir Krivchenkov has pointed out that energy of the ground state and the first excited states for the quantum 3-body problem can be calculated by hand. But in classical mechanics, the analogous problem is less tractable because of chaos.

57 Quoted in Arthur Koestler, *The Sleepwalkers*, Penguin Books, 1990, page 338.

58 An animation and further information can be found at: http://www.scholarpedia.org/article/N-body_choreographies

59 Named after the Earl of Orrery, to whom one was presented in 1704.

60 More formally, this is called the Liapunov time.

61 There is a variant that integrates $1/\log t$ from 2 to x, rather than from 0 to x. This avoids a technical difficulty at $t = 0$, where $\log t$ is not defined. Sometimes the notation $\text{Li}(x)$ is used for this variant, and the function defined in the text is called $\text{li}(x)$.

62 The name 'Pafnuty' is unusual. It led Philip Davis to write a quirky but gripping book: *The Thread: a Mathematical Yarn*, Harvester Press, 1983.

63 This follows from Riemann's curious formula

$$\zeta(1-s) = 2^{1-s}\pi^{-s}\sin\left(\frac{\pi(1-s)}{2}\right)\Gamma(s)\zeta(s)$$

where $\Gamma(s)$ is a classical function called the gamma function, defined for all complex s. The right-hand side is defined when the real part of s is greater than 1.

64 Bernhard Riemann, Über die Anzahl der Primzahlen unter einer gegebenen Grösse, *Monatsberichte der Königlich Preußischen Akademie der Wissenschaften zu Berlin*, November 1859.

65 Riemann defined a closely related function

$$\Pi(x) = \pi(x) + \frac{1}{2}\pi(x^{1/2}) + \frac{1}{3}\pi(x^{1/3}) + \frac{1}{4}\pi(x^{1/4}) + \cdots$$

which counts prime powers rather than primes. From this we can recover $\pi(x)$. Then he proved an exact formula for this modified function in terms of logarithmic integrals and a related integral:

$$\Pi(x) = \text{Li}(x) - \sum_{\rho}\text{Li}(x^{\rho}) + \int_x^{\infty}\frac{dt}{t(t^2-1)\log t}$$

Here Σ indicates a sum over all numbers ρ for which $\zeta(\rho)$ is zero, excluding negative even integers.

66 For example, $x + \sqrt{x}$ is asymptotic to x: the ratio is

$$(x + \sqrt{x})/x = 1 + 1/\sqrt{x}$$

As x becomes large, so does $\sqrt{x}$, so $1/\sqrt{x}$ tends to 0 and the ratio tends to 1. But the difference is $\sqrt{x}$, and that becomes larger and larger as x increases. For instance, when x is 1 trillion, $\sqrt{x}$ is 1 million.

67 Euler's constant is the limit, as n tends to infinity, of

$$1 + \frac{1}{2} + \frac{1}{3} + \cdots + \frac{1}{n} - \log n$$

68 Douglas A. Stoll and Patrick Demichel, The impact of $\zeta(s)$ complex zeros on $\pi(x)$ for $x < 10^{10^{13}}$, *Mathematics of Computation* **276** (2011) 2381–2394.

69 http://empslocal.ex.ac.uk/people/staff/mrwatkin/zeta/RHproofs.htm

70 J. Brian Conrey and Xian-Jin Li, A note on some positivity
conditions related to zeta- and L-functions:
http://arxiv.org/abs/math.NT/9812166

71 The unit 3-sphere comprises all points with coordinates (x, y, z, w)
such that $x^2 + y^2 + z^2 + w^2 = 1$. There are several ways to make the
3-sphere more intuitive. They can all be understood by analogy with
a 2-sphere, and checked using coordinate geometry. One such
description ('solid ball with all surface point identified') is given in
the text, and Figure 57 shows another. To set up the analogy,
observe that if we cut a 2-sphere along its equator we get two
hemispheres. Each flattens out into a disc, and this is a continuous
deformation. To reconstruct the 2-sphere, we just identify
corresponding points on the boundaries of these two discs. In a
sense, we have made a map of the 2-sphere using two flat discs,
much as mapmakers create flat projections of our round planet. We
can construct a 3-sphere using an analogous procedure. Take two
solid balls and identify corresponding points on their surfaces. Now
both have the same surface (because we identified the two surfaces),
and it is a 2-sphere. It forms the 'equator' of the 3-sphere.

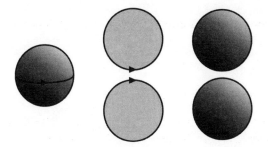

Fig 57 How to make a 3-sphere. *Left:* Cutting a 2-sphere into hemispheres. *Middle:*
Reconstructing the 2-sphere from the two halves by gluing the edges. *Right:* By analogy,
conceptually glue the surfaces of two balls together so that corresponding points are
considered to be identical. This gives a 3-sphere.

72 The usual convention is that we talk of addition and use the
notation $a + b$ when the commutative law is valid, but talk of
multiplication and use the notation ab when it might not be. I have
ignored this convention here because this isn't a textbook on group
theory and 'addition' seems more natural.

73 Start the count at zero. Every time you pass the bus stop going anticlockwise, increase the count by 1; every time you pass it going anticlockwise, decrease the count by 1. At the end of the trip, add 1 if you arrived going anticlockwise, subtract 1 if you arrived going clockwise. The final count is the total number of times you went round the circle, measured in the anticlockwise direction.

74 Stirling's formula states that $n!$ is approximately $\sqrt{2\pi n}(n/e)^n$.

75 William J. Cook, *In Pursuit of the Travelling Salesman*, Princeton University Press, Princeton, 2012. For current information, see http://www.tsp.gatech.edu/index.html

76 Richard M. Karp, Reducibility among combinatorial problems, in R.E. Miller and J.W. Thatcher (eds.) *Complexity of Computer Computations*, Plenum, 1972, pages 85–103.

77 Z. Xia, The existence of noncollision singularities in Newtonian systems, *Annals of Mathematics* 135 (1992) 411–468.

78 http://www.claymath.org/millennium/Navier-Stokes_Equations/

79 See Ian Stewart, *Seventeen Equations that Changed the World*, Profile, 2012, chapter 14.

80 Leonardo Pisano Fibonacci, *The Book of Squares*, annotated and translated by L.E. Sigler, Academic Press, 1987.

81 Leonardo found a family of solutions

$$\left(\frac{m^2+n^2}{2}\right)^2 - mn(m^2-n^2) = \left(\frac{m^2-2mn-n^2}{2}\right)^2$$

$$\left(\frac{m^2+n^2}{2}\right)^2 + mn(m^2-n^2) = \left(\frac{m^2+2mn-n^2}{2}\right)^2$$

where m, n are both odd. The role of d here is played by the number $mn(m^2-n^2)$, and x is $m^2+n^2/2$. Choosing $m = 5$, $n = 4$ leads to $mn(m^2-n^2) = 720$. Moreover, $720 = 5 \times 12^2$. Dividing x by 12 yields the answer.

82 If $x - n$, x, and $x + n$ are squares, then so is their product, which is $x^3 - n^2x$. Therefore the equation $y^2 = x^3 - n^2x$ has a rational solution. Moreover, y is not zero, otherwise $x = n$ so both x and $2x$ are squares, which is impossible since $\sqrt{2}$ is irrational.

Conversely, if x and y satisfy the cubic equation and y is not 0, then

$a = (x^2 - n^2)/y, b = 2nx/y$, and $c = (x^2 + n^2)/y$ satisfy the equations $a^2 + b^2 = c^2$ and $ab/2 = n$.

83 That is,

$$\prod_{p \leqslant x} \frac{N_p}{p} \approx C(\log r)^r$$

where r is the rank, C is a constant, and $\approx$ means that the ratio of the two sides tends to 1 as x tends to infinity.

84 The most likely reason is that these are the natural translations from the languages used by the most prominent mathematicians in the two areas.

85 Why b wasn't the number of bananas, I'm not sure. Perhaps because in post-war Britain, bananas were exotic items seldom seen in the shops?

86 Hence a standard mathematicians' joke. A biologist, a statistician, and a mathematician are sitting outside a café watching the world go by. A man and a woman enter a building across the road. Ten minutes later, they come out accompanied by a child. 'They've reproduced,' says the biologist. 'No,' says the statistician. 'It's an observational error. On average, two and a half people went each way.' 'No, no, no,' says the mathematician. 'It's perfectly obvious. If someone goes in now, the building will be empty.'

87 Bohr may have had a serious point. Scientific theories are tested through their predictions, but few of these foretell the future. Most are if/then statements: if you pass light through a prism it will split into colours. The 'prediction' doesn't say when this will happen. So, paradoxically, we can make predictions about the weather without predicting the weather. 'If the warm air from a cyclone encounters cold air then it will snow' is a scientific prediction, but not a forecast.

88 The quotation or a close variant has been attributed to about thirty different sources, including Sam Goldwyn, Woody Allen, Winston Churchill, and Confucius. See http://www.larry.denenberg.com/predictions.html

89 For the latest information, see the Prime Pages: http://primes.utm.edu.

90 Ilia Krasikov and Jeffrey C. Lagarias, Bounds for the $3x + 1$ problem using difference inequalities, *Acta Arithmetica* **109** (2003) 237–258.

91 Jorge F. Sawyer and Clifford A. Reiter, Perfect parallelepipeds exist. arXiv:0907.0220 (2009).

92 R. Fulek and J. Pach, A computational approach to Conway's thrackle conjecture, *Computational Geometry* **44** (2011) 345–355.

93 http://en.wikipedia.org/wiki/Langton%27s_ant

94 The ABC conjecture states: For any $\varepsilon > 0$ there exists a constant $k_\varepsilon > 0$ such that if a, b, and c are positive integers having no common factor greater than 1, and $a + b = c$, then $c \leqslant k_\varepsilon P^{1+\varepsilon}$, where P is the product of all distinct primes dividing abc.

95 Andrew Granville and Thomas J. Tucker. It's as easy as *abc*, *Notices of the American Mathematical Society* **49** (2002) 1224–1231.
    In September 2012 Shinichi Mochizuki announced that he had proved the ABC conjecture using a radical new approach to the foundations of algebraic geometry. Experts are now checking his 500-page proof, but this may take a long time.

Index